THE MASTERS ATHLETE

Masters Athletes are those that continue to train and compete, typically at a high level, beyond the age of 35 and into middle and old age. As populations in the industrialized world get older and governments become increasingly keen to promote healthy aging and non-pharmacological interventions, the study of Masters Athletes enables us to better understand the benefits of, and motivations for, life-long involvement in physical activity. This is the first book to draw together current research on Masters Athletes.

The book examines the evidence that cognitive skills, motor skills and physiological capabilities can be maintained at a high level with advancing age, and that age related decline is slowed in athletes that continue to train and compete in their later years. Including contributions from leading international experts in physiology, motor behaviour, psychology, gerontology, and medicine, the book explores key issues such as:

- motivation for involvement in sport and physical activity across the lifespan;
- evidence of lower incidence of cardiovascular disease, hypertension, and diabetes;
- the maintenance of performance with age.

Challenging conventional views of old age, and with important implications for policy and future research, this book is essential reading for students and practitioners working in sport and exercise science, aging and public health, human development, and related disciplines.

Joseph Baker is an associate professor in the School of Kinesiology and Health Science at York University in Toronto, Canada. He is the current president of the Canadian Society for Psychomotor Learning and Sport Psychology.

Sean Horton is an assistant professor at the University of Windsor. His research is focused on skill acquisition and expert performance throughout the lifespan, as well as how stereotypes of aging affect seniors' participation in exercise.

Patricia Weir has been a faculty member at the University of Windsor since 1991. Her research interests include the effects of aging on goal-directed movement, psychosocial changes in Masters Athletes, and the role that physical activity plays in developing successful aging.

THE MASTERS ATHLETE

UNDERSTANDING THE ROLE OF SPORT AND EXERCISE IN OPTIMIZING AGING

EDITED BY JOSEPH BAKER, SEAN HORTON, AND PATRICIA WEIR

Routledge
Taylor & Francis Group

LONDON AND NEW YORK

First published 2010
by Routledge
2 Park Square, Milton Park, Abingdon, Oxon, OX14 4RN

Simultaneously published in the USA and Canada
by Routledge
270 Madison Avenue, New York, NY 10016

Routledge is an imprint of the Taylor & Francis Group, an Informa business

Typeset in Zapf Humanist and Eras by
Florence Production Ltd, Stoodleigh, Devon
Printed and bound in Great Britain by
TJ International Ltd, Padstow, Cornwall

British Library Cataloguing in Publication Data
A catalogue record for this book is available from the British Library

Library of Congress Cataloging in Publication Data
The masters athlete: understanding the role of exercise in optimizing
aging/edited by Joseph Baker, Sean Horton and Patricia Weir.
    p. cm.
    Includes index.
    1. Sports for older people.   2. Physical fitness for older people.
    I. Baker, Joseph, 1969–.   II. Horton, Sean.   III. Weir, Patricia.
    GV708.5.M37 2010
    796'.0846 — dc22   2009003722

ISBN10: 0–415–47656–9 (hbk)
ISBN10: 0–415–47657–7 (pbk)
ISBN10: 0–203–88551–1 (ebk)

ISBN13: 978–0–415–47656–0 (hbk)
ISBN13: 978–0–415–47657–7 (pbk)
ISBN13: 978–0–203–88551–2 (ebk)

Father Time is not always a hard parent, and, though he tarries for none of his children, often lays his hand lightly upon those who have used him well; making them old men and women inexorably enough, but leaving their hearts and spirits young and in full vigor.

Charles Dickens, *Barnaby Rudge*

CONTENTS

FIGURES

TABLES

ACKNOWLEDGMENTS

The editors would like to thank Janet Starkes for her feedback during the creation of this text, and Jane Logan for her assistance with the text editing.

PREFACE

The benefits of lifelong involvement in physical activity are well known. They include decreased risk of cardiovascular disease, hypertension, and diabetes (Katzmarzyk et al., 2003), as well as increased physical and mental health (Mazzeo et al., 1998). Despite these benefits, rates of physical activity typically decline with advancing age. Investigations of physical activity involvement across the lifespan show a trend toward peak involvement during early to mid adolescence, followed by decreasing involvement from that point forward (Crocker & Faulkner, 1999; De Knop et al., 1996).

This pattern has important long-term effects. Indeed, much of the decline in physical and cognitive abilities with advancing age is thought to be the result of disuse rather than age per se (Maharam et al., 1999). Studies of cognitive and motor skills, such as chess (Charness, 1981) and typing (Salthouse, 1984), as well as physiological capacities, such as maximal strength (Tarpenning et al., 2004), suggest performance can be maintained at high levels in spite of advancing age, provided there is continued involvement in the activity. As a result, the lack of physical activity in older adults has been identified as a primary contributor to decreases in functional capacity and increases in morbidity and mortality (DiPietro, 2001).

One group that deviates from the typical profile of aging and the corresponding decline in physical activity levels is Masters Athletes. These athletes typically maintain higher-than-average levels of physical activity throughout the lifespan (Hawkins et al., 2003) and are unique because they continue to physically train and compete well into old age. Compare this with Canadian statistics that show, by the age of 50, only one in ten individuals is motivated to be involved in sport activities at least once per week (Sport Canada, 2003). Continued involvement in sports has its benefits. Sport scientists (e.g., Starkes et al., 1999) have suggested that prolonged training by Master Athletes plays a critical role

1

in the maintenance of athletic performance even in the face of predicted age-related decline. The physiological changes that occur with age are well documented — age changes for maximal heart rate (Hagberg et al., 1985) and aerobic capacities (Eskurza et al., 2002; Hawkins et al., 2001; Pimentel et al., 2003) are significant. Yet age-related physiological decline is not as severe in Masters Athletes.

The number of older athletes is greater than ever before, and all of the evidence to date illustrates that Masters Athletes are the physical elite and 'best preserved' of their age cohorts. As a result, some (Hawkins et al., 2003) have suggested they represent a model of 'successful' aging, and researchers have begun utilizing this population to examine a host of issues relative to aging, physical/cognitive functioning, and health.

This book brings together leading researchers from around the world to discuss the most recent research and its intriguing implications for both aging athletes and the population as a whole. In addition, the authors have identified areas that require further inquiry — research questions that will form the basis for future work with this important population. In general, this text is divided into four sections. Section One provides a summary of some of the most pertinent issues in the field (Chapter 1) and the statistical methods used to evaluate age-related declines in performance (Chapter 2). Section Two summarizes research on the effect of aging on muscle recovery from exercise (Chapter 3) and cardio-respiratory adaptations with age (Chapter 4). Chapter 5 summarizes research showing a high degree of performance maintenance in highly skilled groups, and Chapter 6 considers how age affects recovery from training stress (among other things). Section Three focuses on psychosocial issues in Masters sport, covering topics ranging from the development and maintenance of motivation (Chapter 7) to the role that Masters Athletes play in challenging some of the negative stereotypes of aging that exist in society (Chapter 8), and how Masters sport might assist an individual's navigation through the aging process (Chapter 9). In Section Four, the book considers some of the larger issues in public health. Chapter 10 examines Masters Athletes as they relate to theories of 'successful aging', while Chapter 11 examines the epidemiology of injury in this population. Finally, Chapter 12 provides a critique of the book with specific attention to limitations in current knowledge and key directions for future work.

Perhaps the greatest advantage of a book of this nature is the possibility for cross-fertilization of ideas between researchers from different domains. This text summarizes current research from the fields of medicine, physiology, motor behavior, psychology, and gerontology, and reinforces the value of Masters Athletes as a research population for examining issues related to optimal and

2

successful aging. Considering the demographic trends in many industrialized countries of the world, more attention to the issue of healthy and successful aging is clearly warranted.

REFERENCES

Charness, N. (1981). Search in chess: Age and skill differences. *Journal of Experimental Psychology: Human Perception and Performance, 7*, 467–476.

Crocker, P.R.E., & Faulkner, R.A. (1999). Self-report of physical activity intensity in youth: Gender and grade considerations. *AVANTE, 5*, 43–51.

De Knop, P., Engstrom, L-M., Skirstad, P., & Weiss, M. (1996). *Worldwide trends in youth sport*. Champaign, IL: Human Kinetics.

DiPietro, L. (2001). Physical activity in aging: Changes in patterns and their relationship to health and function. *Journal of Gerontology: Medical Sciences, 56*, Special 2, 13–22.

Eskurza, I., Donato, A.J., Moreau, K.L., Seals, D.R., & Tanaka, H. (2002). Changes in maximal aerobic capacity with age in endurance-trained women: 7-yr. follow-up. *Journal of Applied Physiology, 92*, 2303–2308.

Hagberg, J.M., Allen, W.K., Seals, D.R., Hurley, B.F., Ehsani, A.A., & Holloszy, J.O. (1985). A hemodynamic comparison of young and older endurance athletes during exercise. *Journal of Applied Physiology, 58*, 2041–2046.

Hawkins, S.A., Marcell, T.J., Jaque, V., & Wiswell, R.A. (2001). A longitudinal assessment of change in VO_2max and maximal heart rate in master athletes. *Medicine and Science in Sports and Exercise, 33*(10), 1744–1750.

Hawkins, S.A., Wiswell, R.A., & Marcell, T.J. (2003). Exercise and the master athlete: A model of successful aging? *Journal of Gerontology: Medical Sciences, 58A*, 1009–1011.

Katzmarzyk, P.T., Janssen, I., & Ardern, C.I. (2003). Physical inactivity, excess adiposity and premature mortality. *Obesity Reviews, 4*, 257–290.

Maharam, L.G., Bauman, P.A., Kalman, D., Skolnik, H., & Perle, S.M. (1999). Masters athletes: Factors affecting performance. *Sports Medicine, 28*, 273–285.

Mazzeo, R.S., Cavanagh, P., Evans, W.J., Fiatarone, M., Hagberg, J., McAuley, E., & Startzell, J.K. (1998). Exercise and physical activity for older adults. *Medicine & Science in Sports & Exercise, 30*, 1–13.

Pimentel, A. E., Gentile, C.L., Tanaka, H., Seals, D.R., & Gates, P.E. (2003). Greater rate of decline in maximal aerobic capacity with age in endurance-trained than in sedentary men. *Journal of Applied Physiology, 94*, 2406–2413.

Salthouse, T. (1984). Effects of age and skill in typing. *Journal of Experimental Psychology: General, 113*, 345–371.

Sport Canada. (2003, May). Sport participation in Canada: 1998 report. Retrieved from http://www.canadianheritage.gc.ca/progs/sc/psc-spc/index_e.cfm.

Starkes, J.L., Weir, P.L., Singh, P., Hodges, N.J., & Kerr, T. (1999). Aging and the retention of sport expertise. *International Journal of Sport Psychology, 30*, 283–301.

Tarpenning, K.M., Hamilton-Wessler, M., Wiswell, R.A., & Hawkins, S.A. (2004). Endurance training delays age of decline in leg strength and muscle morphology. *Medicine and Science in Sports and Exercise, 36*, 74–78.

SECTION ONE

INTRODUCTION TO MASTERS SPORT AND THE STUDY OF OLDER ATHLETES

CHAPTER ONE

THE EMERGENCE OF MASTERS SPORT

Participatory trends and historical developments

PATRICIA WEIR, JOSEPH BAKER, AND SEAN HORTON

> We are aging — not just as individuals or communities but as a world. In 2006, almost 500 million people worldwide were 65 and older. By 2030, that total is projected to increase to 1 billion — one in every eight of the earth's inhabitants. Significantly, the rapid increases in the 65-and-older population are occurring in developing countries, which will see a jump of 140 percent by 2030.
>
> US Department of State, April 2007

Global population aging is a function of two factors: decreased fertility rates and improvements in health and longevity. Until the mid-1960s, the fertility rate in Canada was equal to three children or more per woman. Since that time, the fertility rate has experienced a rapid decline, sitting below the rate for natural replacement of the population for the last 30 years (Health Canada, 2002). Similar trends exist in many westernized countries, and, most surprisingly, this trend is seen in 44 per cent of less developed nations. The demographics of the global population will continue to change. The United Nations estimates that in 2017, the percentage of the population over 65 years of age will exceed the percentage of the population under five years of age, a shift that is expected to continue for many decades to come (United Nations, 2005).

In Canada, as the baby boomers (those born between 1946 and 1964) age, the population of seniors is expected to grow to 6.7 million in 2021 and 9.2 million in 2041. By 2041, one in four Canadians will be a senior. Over the next four decades, the growth of the senior population will account for almost half the population growth in Canada (Health Canada, 2002). In Canada, and around the world, the fastest growing segment of the older population is the 'oldest-old', or seniors aged 85+ years. Currently the oldest-old make up seven per cent of the world's population over 65 years. More developed countries have approximately ten per cent of the seniors in the oldest-old age group, while less developed countries have approximately five per cent. The majority of the world's oldest-old live in six countries: in descending order, China, the United States, Japan, India, Germany, and Russia.

With an aging population comes a whole host of new challenges. Issues related to health and well-being, retirement, economic sustainability, and changes in family structure all take on new importance. Specific to this book is the health and well-being of the world's senior population. Chronic diseases will increase disproportionately given the rapid aging of the oldest-old and will impact the health care system in every country. While chronic conditions are disabling, costly, and cause limitations in activity, they are also the most preventable.

Research extolling the importance of physical activity for improving and maintaining health has encouraged regular physical activity in persons of all ages. Although this message has not been adopted by the vast majority, particularly in older age groups, a significant minority maintains involvement in vigorous physical activity throughout the lifespan. This highly active cohort has been the catalyst for significant change in the organizational structure of sport. 'Masters' sport evolved out of elite competitive sport as a means of continuing participation for athletes who are past the typical age of peak performance. Usually Masters competition is organized into five- or ten-year age groupings (40–44, 45–49, etc.) starting from 30 or 35 years of age, although this can vary significantly by sport and competition. For instance, in New York's 2008 Empire State Games, anyone over the age of 22 was allowed to compete in the Masters gymnastics competition, but participants in bowling, archery, and fencing had to be over the age of 50.

HISTORICAL DEVELOPMENT OF MASTERS SPORT

Although it is difficult to identify the specific 'birthdate' of Masters sport, it is generally accepted that its origins were in the mid-1960s. In the United States, Masters track and field can be traced to David Pain, an attorney and runner from La Jolla, California, who created the first 'Masters Mile' in 1966. The concept of a competitive venue for athletes who were 'past their prime' proved to be very popular, and the concept was quickly expanded to track and field meets completely restricted to Masters Athletes (Wallace, 1991). The first Masters US Track and Field Championships were held in 1968 and included 130 competitors (all men; women were not included in these events until 1971). Olsen's *Masters track and field: A history* (2000) suggests that Pain's trips abroad in 1971 and 1973 laid the groundwork for Masters level competition in Europe and Australasia

8

respectively. The first World Masters Championships took place in Toronto, Canada, in 1975 and included 1,400 competitors.

Similarly, Masters swimming started as a 'one-time' event in the United States that quickly expanded to national and international levels. The first National Masters Swimming Championship was held in Amarillo, Texas, in 1970 with 46 competitors. During this event, the United States Masters Swimming organization was created. Other countries were quick to follow suit; Canada created its first Masters swim club at the University of Toronto in 1971, and Australia created the AUSSI Masters Swimming organization in 1975 (Dionigi, 2008). The first World Masters Swimming Championships were held in Tokyo in 1986. While track and field and swimming have the longest history of competition at the Masters level, most sports now have competitive opportunities for older athletes.

The first World Masters Games (WMG) were held in Toronto in 1985 and included 8,305 participants representing 61 countries and participating in 22 sports. Since this first event, participation in the WMG has increased considerably (see Figure 1.1). According to the website for the 2009 WMG in Sydney, Australia, over 30,000 participants are expected to compete in 28 sports. The 2005 games in Edmonton, Canada, included 21,600 athletes, and registration had to be closed months prior to the games because organizers reached the capacity of the competition venues. Historically, the WMG have been geared primarily to summer sports; however, the first World Masters Winter Games will take place

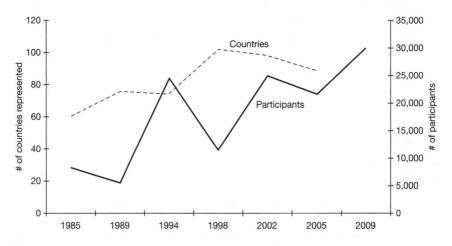

Figure 1.1 Increase in countries represented and competitors participating in World Masters Games since their inception in 1985

Note: numbers for 2009 reflect estimated number of competitors.

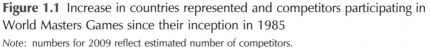

in Bled, Slovenia, in 2010 and will include the core sports of alpine and cross country skiing, biathlon, curling, ice hockey, ski jumping, and speed skating.

The WMG are overseen by the International Masters Games Association, whose goal is to represent Masters sport and 'promote lifelong competition, friendship and understanding between mature sportspeople, regardless of age, gender, race, religion, or sport status'. The organization is made up of members from individual sporting associations and the International Olympic Committee. Although starting out as isolated events in individual sports, Masters-level participation has evolved into a very sophisticated form of competition with a comprehensive organizational structure attending to issues ranging from determining world records to policing doping infractions among participants.

MASTERS ATHLETES AND HUMAN AGING

Masters Athletes represent an intriguing group for researchers due to the fact that they represent some of society's most successful agers, at least from a physical standpoint. While it is clear that physical and cognitive abilities decline with age, there has been considerable debate as to whether the bulk of this decline is actually a result of age, or a result of increasingly sedentary lifestyles. Many researchers (e.g., Maharam et al., 1999) speculate that lifestyle factors are actually responsible for much of the decline that is traditionally attributed to old age.

Researchers have postulated a general rate of performance decline from 0.5 per cent (Bortz & Bortz, 1996) to one per cent annually after peak performance (see Hawkins, Chapter 4), although this tends to vary considerably depending on a number of factors, including frequency and intensity of training and the domain in which an individual is participating. While Hawkins focuses on physiological performance measures, specifically VO_2max, examinations of expert performance in areas that rely more heavily on cognitive skills (e.g., golf; see Chapter 5 by Baker & Schorer) have found rates of performance decline to be considerably less than 0.5 per cent.

Masters Athletes who engage in high levels of training represent the upper levels of physical performance, thereby helping to control the 'disuse' factor. Yet even in these highly trained individuals, measuring performance decline still tends to be confounded by a decrease in weekly training schedules and an increase in body fat (see Shephard, Chapter 12). Gaining precise insights into the rates of aging is further complicated by the fact that the number of competitors decreases in the oldest age groups (Figure 1.2), and that Masters games competitors are mostly white and well educated, thus limiting

10

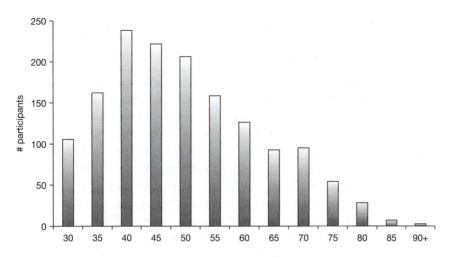

Figure 1.2 Number of participants by age in athletic events at the 2005 World Masters Games, Edmonton

generalizability. Moreover, women are underrepresented in a number of events, perhaps due to lingering societal stereotypes, particularly of older women partaking in certain sporting activities (see Tanaka, Chapter 3). Thus, research into performance decline may reflect important social factors along with biological constraints.

Even determining the age of peak performance has been difficult to pinpoint, and may reflect social, along with physiological and biological considerations. There have been attempts — notably by Lehman (1953) and by Schulz and Curnow (1988) — to determine the age of peak performance across a variety of domains. Schulz and Curnow used archival records from the Olympic Games that showed, for example, that female swimmers tend to peak at age 17 and male swimmers at 19. (Stones, in Chapter 2, shows that the mean age of those setting swimming records in 2008 was early 20s.) That does, however, make it difficult to explain the triple–silver medal performance of 41-year-old Dara Torres in the 2008 Games.

While there were many intriguing themes that emerged during the 2008 Beijing Games — notably the issue of human rights in the host country, and the youthfulness of their gold-medal-winning gymnasts — the relatively advanced age of some competitors also attracted considerable attention. Thirty-three-year-old Oksana Chusovitina, competing in her fifth Olympics, astounded the gymnastics community by taking silver in the vault (recall that gymnasts become eligible for Masters competitions at age 22). Constantina Tomescu-Dita, a

38-year-old mother, won the women's marathon. Jujie Luan, age 50, represented Canada in fencing. She returned to Beijing to a hero's welcome, having won gold for her country of birth in 1984, China's first gold in that sport.

Perhaps most remarkable, however, was Torres, who missed out on gold in the 50m freestyle swimming event by 1/100th of a second. She became the oldest medalist in the history of Olympic swimming, a record that had belonged to William Robinson, who was 38 when he won silver 100 year ago, in 1908 (Arthur, 2008). As Arthur notes, Torres' accomplishments have aroused suspicions, something she has tried to counter by submitting to voluntary third-party drug testing and supplying samples for future testing when technology improves. Torres' feats do seem superhuman, particularly considering that she recently became a mother, has endured surgery on her shoulder and knee, and has been diagnosed as asthmatic (Arthur, 2008).

This was Torres' fifth Olympic Games; she had originally retired at age 25, believing that she was too old, which is certainly what some of the academic literature suggested. She made her first comeback for the Sydney Games in 2000, and came back most recently in Beijing. Of particular interest is that in 2006, after giving birth to her daughter, she entered Masters level events. After posting times that were internationally competitive, Torres was emboldened to try yet another comeback.

With Torres' swim times improving into her 40s — her silver-medal time in the 50m freestyle was a personal best — interesting questions are raised about peak performance and the inevitability of age-related decline. Since data on peak performance is often based on archival records (i.e., Schulz & Curnow, 1988), there are likely important social and historical factors affecting the results. Athletes who retire due to the perception that they are 'too old' is one such consideration.

Financial constraints are another important factor. David Ford, age 37, is a kayaker who garnered a sixth-place finish in Beijing. Ford invested $80,000 of his own money into his training after his funding was cut because he was told, 'I was too old and my sport wasn't relevant in Canadian culture' (Christie, 2008). Ford intends to compete again in London in 2012. The reality is, however, that funding decisions by national bodies, or coaching decisions that favor younger athletes, may reflect an age bias that ultimately affects our conceptions of peak performance.

It is likely that age limits will be further challenged in years ahead as people take inspiration from athletes such as Torres, Tomescu-Dita, and Luan, although it is difficult to determine how many middle-aged mothers will leap back in the pool or dust off old running shoes based on their examples. There are certainly

12

numerous anecdotal reports of Olympic athletes inspiring others to get involved in sport, and a variety of athletic clubs — from gymnastics to diving to trampoline — generally see a spike in registrations immediately following the Olympic Games (Mick, 2008). Whether this translates into sustained engagement and increased overall societal involvement in sport remains debatable. Hogan and Norton (2000) attempted to determine how the Australian Institute of Sport, created in 1981, has fared in its twin objectives of 1) excellence in sport performance and 2) increased participation in sports and sports activities. The authors concluded that, while increased funding at the elite level has translated into a greater medal tally for Australia, the effect on mass participation rates was more equivocal.

This does raise important public policy questions, particularly if part of the rationale for funding elite-level sports is that it ultimately translates into greater sport- and physical-activity participation by the general populace. While athletes like Torres are held up as role models, the extent to which they will inspire behavior change on a grander scale is an open question. Similarly, 76-year-old marathoner Ed Whitlock has been extensively profiled in the popular media due to his remarkable performances and his extensive training regimen (see Horton, Chapter 8). Just as Torres, Jujie Luan, and Constantina Tomescu-Dita challenged our notions of what it means to be a middle-aged mother, Whitlock defies many of the popular stereotypes of aging that we hold in our society. Research into whether these athletes can have any meaningful impact on physical activity levels of the population as a whole, however, is in its very early stages.

It appears that peak performance and rates of aging are an intriguing mix of a number of different variables. This is reinforced by the fact that there continue to be improvements in age-class records (Stones, Chapter 2). Indeed, results from the New York marathon suggest that there is greater performance improvement by older Masters groups than by younger athletes (Jokl et al., 2004). What does seem clear is that we will continue to be surprised in coming years, as athletes like Torres, Whitlock, Luan, and others force us to re-examine our notions of performance and aging.

REFERENCES

Arthur, B. (2008, Aug 10). Torres a winner in a suspicious age. *National Post.* Retrieved from http://www.nationalpost.com/sports/beijing-games/story.html?id=714251.

Bortz, W.M., & Bortz, W.M. (1996). How fast do we age? Exercise performance over time as a biomarker. *Journal of Gerontology: Medical Sciences*, 51A, M223–M225.

Christie, J. (2008, Aug 12). Ford 6th in kayak final. *Globe and Mail*. Retrieved from http://www.globesports.com/servlet/story/RTGAM.20080812.wolymford finals12/BNStory/beijing2008/home.

Dionigi, R. (2008). *Competing for life: Older people, sport and ageing*. Saarbrüecken: VDM Verlag Dr. Müller.

Health Canada (2002). *Canada's aging population*. (Cat. H39–608/2002E). Ottawa, Ontario: Health Canada.

Hogan, K., & Norton, K. (2000). The 'price' of Olympic gold. *Journal of Science and Medicine in Sport, 3,* 203–218.

Jokl, P., Sethi, P.M., & Cooper, A.J. (2004). Master's performance in the New York City Marathon, 1983–1999. *Sports Medicine, 35,* 1017–1024.

Lehman, H.C. (1953). *Age and achievement*. Princeton, New Jersey: American Philosophical Society.

Maharam, L.G., Bauman, P.A., Kalman, D., Skolnik, H., & Perle, S.M. (1999). Masters athletes: Factors affecting performance. *Sports Medicine, 28,* 273–285.

Mick, H. (2008, Aug 15). Swim like Phelps, paddle like van Koeverden. *Globe and Mail*. Retrieved from http://www.theglobeandmail.com/servlet/story/RTGAM.20080815.wleffect15/BNStory/lifeMain.

Olson, L.T. (2000). *Masters track and field: A history*. Jefferson, North Carolina: McFarland.

Schulz, R., & Curnow, C. (1988). Peak performance and age among superathletes: Track and field, swimming, baseball, tennis, and golf. *Journal of Gerontology: Psychological Sciences, 43,* P113–120.

United Nations (2005). *World population prospects. The 2004 revision.* United Nations Department of Economic and Social Affairs, Population Division.

US Department of State (2007). *Why population aging matters: A global perspective.*

Wallace, L. (1991). *Oral history: Women in Masters track and field.* Unpublished Masters thesis. Central Washington University.

14

CHAPTER TWO

STATISTICAL MODELING OF AGE TRENDS IN MASTERS ATHLETES

MICHAEL STONES

Research on age trends in Masters Athletes always seemed special to me for several reasons: high quality of measurement on familiar and highly practiced activities; performance trends not confounded by effects of chronic incapacity or physical inactivity; a rare opportunity to study expertise at the highest level. As someone involved in such research since near its inception in the 1970s, I'm delighted with the opportunity to revisit old discoveries, trace their fate over the ensuing decades, and try to introduce new methodology and findings.

A brief reminiscence might be a good way to begin because the process of discovery has relevance to its outcomes. Three major changes from my earliest research until now are significant: these relate to accessibility of data, quality of record performances, and statistical methodology. First, records then were not easily accessible; one compilation I tracked down was available only on mimeographed sheets. International bodies such as World Masters Athletics (WMA) and the Fédération International de Natation (FINA) now put world records for track and field and swimming on the Internet for anyone to peruse. Second, the quality of performance by Masters Athletes is much higher now than it was then because of greater participation and more opportunities for competition. Third, for analysis then I used a HP55 calculator that was high-tech for the era. It had four built-in curve-fitting models that seemed so sophisticated. Statistical procedures such as mixed linear analysis, essential for analysis of nested data, were not even a dream in some developer's eye. Consequently, analysis now compared with then should be more sophisticated and based on more readily accessible data of higher quality.

It is for such reasons that the first section of this chapter is a historical overview that traces the fate of early findings in subsequent replication. It is a testament to the robustness of the early research that successful replication proved the rule. This section also relates the early research to the intellectual climate of the time in order to illustrate influences on direction and perceived significance. The second section examines interpretative models that found favor in different eras. A key difference concerns beliefs about the continuity or discontinuity of aging effects throughout adulthood. The third section reports new findings that extend this discussion to the age of peak performance in young athletes.

Although the aforementioned sections contain discussion of methodology, the approach is descriptive rather than evaluative. The fourth section contains critical evaluation of statistical models, adaptations to augment fit to data, and nesting as property of such data. The fifth section reports new findings with track and field records that take account of nesting. Finally, the sixth section draws overall conclusions.

HISTORICAL OVERVIEW

Statistical analysis of age trends in sports records became part of my research after I happened upon a copy of *Runner's World* magazine in 1979. That particular issue contained what was probably the first compilation of age-class running records ever published in a popular national magazine (Mundle & Brieger, 1979). What became known as 'Masters athletics' was new at that time, with the first world championships held in Toronto only four years earlier. So I was excited that these records might be a key to unlocking some hitherto unknown secrets of aging among physically elite older people.

The period when this happened was pivotal in the history of gerontology, marking a transition from near-universal acceptance of the irreversible decrement model of aging to its replacement by a decrement-with-compensation model that gave rise to concepts such as 'successful' aging (Rowe & Kahn, 1987; see Weir, Chapter 10). Research design was likewise in a transitional phase, with attention newly directed toward performance by experts rather than novices. Research on older elite athletes complemented these paradigm shifts particularly because of findings that physical disuse contributed as much as aging effects as irreversible physiological and psychomotor declines (Cooper, 1977; Shephard, 1978; Smith & Gilligan, 1983; Spirduso, 1980). Consequently, that fortuitous reading of *Runner's World* happened at an apposite time.

Running records

The study that emerged (Stones & Kozma, 1980) was neither the first analysis of running records nor the first to analyze age trends in those records. The former

16

michael stones

distinction belongs to Henry (1955) who attempted to predict world records over distances from 60 yards to the marathon. Henry also made an important point with respect to measurement quality: namely, that running records provide data gathered under conditions controlled as rigorously, or more rigorously, than those from any well-controlled field experiment. The first studies of age trends in running records were by Moore (1975) and Salthouse (1976). Moore (1975) applied a complex exponential model to records at four event distances and concluded that, beyond the age of peak performance, there were greater declines with age in shorter events than in longer ones. Salthouse (1976) analyzed records from nine events by comparing ratios of age-class records to peak performances. He failed to find any effect of event distance. Because both studies analyzed records from the same 1974 compilation, their differing conclusions appear to relate to sampling of events or choice of statistical model.

The records we analyzed were at a higher level of performance than those available to Moore (1975) and Salthouse (1976). The performance gains resulted from higher levels of participation, training, and competitive opportunities throughout the intervening pentad (e.g., records by Canadian runners in 1979 showed increments of up to ten per cent compared with those of five years earlier). We also took care to compare analyses of records by multiple statistical models and used an age range within which year-by-year records showed low variation (i.e., ages 40–74 for males). The model providing best fit over distances ranging from the 100 yards to marathon expressed performance time as a power function of distance and an exponential function of age. An equivalent and conceptually easier way to express age trend within this model is that the logarithm of performance time varies linearly with age.

The most surprising finding from the study was that performance declined with age more steeply in longer runs than the sprints. This finding contradicted a widely believed myth that training could compensate for aging effects more in longer runs than the sprints. That myth proved particularly difficult to dispel. More than a decade later, undergraduate gerontology textbooks continued to perpetuate it (e.g., Rybash et al., 1991, pp. 92–95). However, truth prevailed eventually with consistent replication of findings that performance decline is higher in longer runs than sprints (Baker et al., 2003; Fair, 2007a; Young et al., 2008).

Sex differences

Limitations to the Stones and Kozma (1980) study included the analysis of running records only of male athletes. It was possible to analyze sex differences with data from the same compilation of records but with a compromise. The compromise called for restricting the upper limit of age range analyzed to 63 years because the female records beyond that age were highly variable. We decided to accept this compromise.

With the same statistical model used in the initial study, our findings showed steeper performance declines with age for both sexes in longer runs than the sprints, with higher declines for females than males in both categories of event (Stones & Kozma, 1982b). The sex difference obtained in that study has stood the test of time.

Subsequent studies included age ranges from 30 to >90 years, obtained data from different sports, and used different statistical models. All showed higher performance loss with age by females than males. Examples include cross-sectional trend in track and field (Baker et al., 2003), and cross-sectional and longitudinal findings in short distance swimming (Donato et al., 2003; Fair, 2007a; Tanaka & Seals, 1997).

Track and field records

In order to model age trend over an extended array of track and field records, we realized that simple application of an exponential model might not suffice. The reason is that performance was on scales that differ in direction as well as range. Those expressed by time showed an increasing trajectory with age (e.g., running, hurdling, racewalking, steeplechasing); those expressed by distance showed a decreasing trajectory (e.g., jumping, pole vaulting). After examining the fit of different statistical models to performances in 18 events, we opted for a second-order polynomial model that accounted for the most variance in 17 of the 18 events. Because the aim was to compare declines across events with a standardized performance measure, we transformed both record performances and age into standard scores. The data were 1979 world records for all events except the racewalks, for which we used first-place finishing times in the World Masters Track and Field Championships, because no world records were then available (Stones & Kozma, 1981).

The findings again showed greater performance decline with age for longer runs than for the sprints; however, age trends for the other events dramatically changed our theoretical outlook. For short-duration events, performance loss was lower in the sprints than the hurdles and jumps; for long-duration events, performance loss was lower in the racewalks than the runs and steeplechase (Stones & Kozma, 1981). Subsequent research with different statistical models consistently replicated these findings (Baker et al., 2003; Fair, 2007a; Stones

18

michael stones

& Kozma, 1996). A theoretical model proposed to explain the trends related peak power cost to energy demand (Stones & Kozma, 1986a).

Swimming records

The earliest analyses of USA Masters swimming records were by Hartley and Hartley (1984a). Although they concluded that speed of swimming showed greater decline with age in shorter than longer events, we challenged this interpretation in an ensuing debate (Hartley & Hartley, 1984b; Stones & Kozma, 1984a, 1986b). The crux of the matter is that speed is a compound index of distance and time. With speed expressed as a function of age and compared at different event distances, distance is present on both sides of the predictive equation (i.e., Speed = Distance/Time = f[Distance + Age]). Only with such confounding removed can we express performance as independent functions of distance and age.

In the absence of such confounding, reanalysis of Hartley and Hartley's (1984a) data and repeated replication studies showed greater performance loss with age in longer than shorter events (Donato et al., 2003; Fair, 2007a; Stones & Kozma, 1986b, 1996; Tanaka & Seals, 1997). Figure 2.1 illustrates this trend with 2007 FINA short-course world records averaged over sex for the 50m and 1500m freestyle. The measure of performance in this graph is a logarithm of the reciprocal of time normalized to 1.0 at the youngest age level. Such a depiction represents performance as proportionate to that at the youngest age.

Other findings with swimming records show greater decline with age in the butterfly than in any other stroke (Stones & Kozma, 1986a). This finding is consistent over event distance and replicated with recent swimming records (Fair, 2007a; Stones, 2001). Figure 2.2 illustrates the trend, with 2007 FINA short-course world records averaged over sex for the 50–100m. Stones and Kozma (1986a) interpreted these finding as due to higher peak-power cost in the butterfly, as evidenced by oxygen uptake and tethered motion studies (Astrand & Rodahl, 1977, pp. 586–589; Holmér, 1974; Magel, 1970).

Historical and longitudinal trends

The final of our early forays into Masters track and field records included the study of longitudinal trend (Stones & Kozma, 1984a) and comparisons of cross-sectional, longitudinal, and historical trends by leading Canadian male athletes over the period 1972–1979 (Stones & Kozma, 1982a). To compare data using

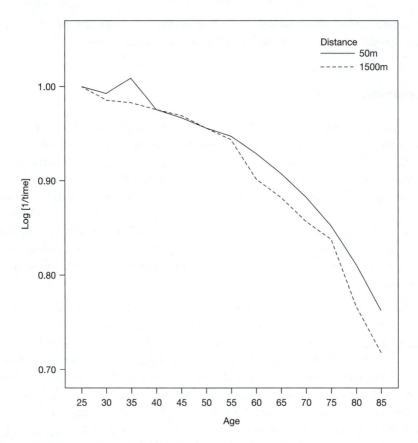

Figure 2.1 Proportionate decline in freestyle swimming performance after age 25 years

a single metric, Stones and Kozma (1982a) used an unbiased estimate of proportionate yearly change computed separately for each data set. We also restricted the events analyzed to those for which at least ten athletes provided at least three records. There were six such events including the long jump and running at distances from 100m to the marathon.

The findings showed mean yearly performance declines of .76 per cent with the longitudinal data, and 1.58 per cent with the cross-sectional data, and mean yearly historical improvements of .32 per cent for athletes aged 40–49 years, and 2.24 per cent for athletes aged 50–74 years (Stones & Kozma, 1982a). Longitudinal declines in performance were clearly lower than cross-sectional declines, with historical improvement mainly present in athletes aged over 50 years.

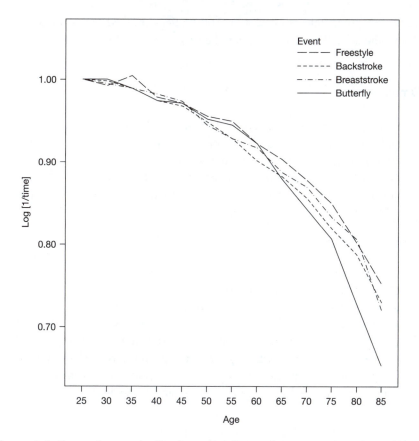

Figure 2.2 Proportionate decline in swimming performance after age 25 years in four swimming strokes at distances 50–100m

Subsequent research that compared longitudinal and cross-sectional performance decline with age used second-order polynomial models with findings of lower quadratic coefficients for longitudinal data in running (Starkes et al., 1999; Young & Starkes, 2005; Young et al., 2008) and swimming (Weir et al., 2002). These findings suggest that the acceleration of performance decline with age is lower with longitudinal than cross-sectional data (i.e., linear rather than accelerated decline). The researchers reasoned that attenuated performance decline with longitudinal data reflects moderating effects due to continued training throughout the longitudinal span.

Research on historical trend also supported the early findings. One such study compared the top 50 finishing times by age group in the New York City marathon

statistical modeling of age trends

from 1983–1999 (Jokl et al., 2004). The findings showed higher performance improvement by the Masters groups than by younger athletes.

Summary

In summary, evidence discussed in this section shows that discoveries made during the first decade of statistically modeling performances by Masters Athletes remain robust and reliable more than a quarter-century later. Despite increases in athletic participation, improvements in age-class records, and changing preferences of statistical model, the following trends remain robust as evidenced by replication in two or more studies.

Performance decline:

- accelerates with age;
- is greater for females than males;
- is greater with cross-sectional than longitudinal data;
- is greater for longer than shorter events within any event category (e.g., longer runs versus sprints; 1500m versus 50m freestyle swims);
- is greater for events with higher peak power costs when duration is comparable (e.g., hurdles and jumps versus sprints; runs versus racewalks; butterfly versus backstroke, breaststroke, and freestyle).

INTERPRETATION OF AGE TRENDS

Attempts to explain these age trends include differential participation, differential training, and bioenergic loss. All three models include examples of what might be termed a discontinuity hypothesis, wherein reasons cited for accelerating performance loss with age are discontinuous from influences on peak performance at younger ages. This section reviews these models and discusses an alternative perspective.

Accelerating performance loss with age could be due to decreasing participation rates. Because the pool of younger athletes (e.g., aged 35–45) is larger than the pool aged 85–95 years, the caliber of records might reflect the size of the pools. Fairbrother (2007) tested this hypothesis in 1500m freestyle swimming but concluded that disproportionate sampling at different ages failed to enhance the prediction of accelerated performance loss at older ages. Differential participation also fails to explain differences in performance decline across events. One such example derives from a comparison of running and racewalking over

michael stones

similar event distances. Participation rates are higher in running than racewalking at all age levels; however, racewalking, rather than running, shows the lower performance decline with age (Baker et al., 2003; Stones & Kozma, 1996). Consequently, support for the differential participation model seems meager at best.

The differential training model proposes that continuous training over many years attenuates acceleration in performance decline with age, although perhaps to a lesser extent for events of prolonged duration (Young & Starkes, 2005; Young et al., 2008). Training patterns also differ with age, such that older athletes train more for endurance than for strength or competition (Weir et al., 2002). Consequently, changes in the quantity and quality of training may contribute to accelerated performance loss among older athletes.

Although this model is consistent with some established age trends, the direction of causality is less assured. Are changes in training a cause of bioenergic loss, or does the latter contribute to changes in training, as Tanaka and Seals (2003, 2008) suggest? Also, proponents of the differential training model have yet to explain differences in age trend among events with similar energy demand (e.g., jumps versus sprints; butterfly versus backstroke, breaststroke, and freestyle; longer runs versus racewalks). Support for this model therefore seems tentative.

It is hard to dispute that bioenergic losses are responsible for declines in athletic performances with age. Tanaka and Seals (2003, 2008) argue that accelerating performance deterioration in advanced age is discontinuous from linear trends found earlier in athletic careers because late-life bioenergic loss involves cost to both performance and the training necessary to maintain performance. The effects of such loss may affect performance in events of longer rather than shorter duration because the costs on training are higher (e.g., Young et al., 2008). On the other hand, continuity models suggest that bioenergic loss has a curvilinear trajectory throughout adulthood, with effects on athletic performance dependent on requirements such as peak power cost and energy demand (Moore, 1975; Salthouse, 1976; Stones & Kozma, 1986b). A study based on one such model raised the question of when such effects first become visible (Stones & Kozma, 1996). The next section describes that study and an attempt at replication with current world records.

AGE AT PEAK PERFORMANCE

Implicit in continuity models is an assumption that cumulative benefits to performance through training and experience are offset by aging effects. We reasoned that, if aging effects are continuous throughout adulthood, they ought to be visible not only in performance declines by aging athletes but also in ages of peak performance by younger athletes (Stones & Kozma, 1996). Figure 2.3 illustrates postulated trajectories for the effects of bioenergic loss in events associated with shallow or steep performance declines with age.

The hypothetical curves in Figure 2.3 show a developmental transition from bioenergic gain to loss at a somewhat arbitrary age of 21 years. If the potential to compensate for bioenergic loss through training and experience relates negatively to the extent of loss, the age range when peak performance is tenable should be wider in events associated with shallow rather than steep performance declines with age. In other words, age at peak performance *can* be older in such events. Conversely, age at peak performance *should* be younger in events with steep performance declines with age. Consequently, this model predicts

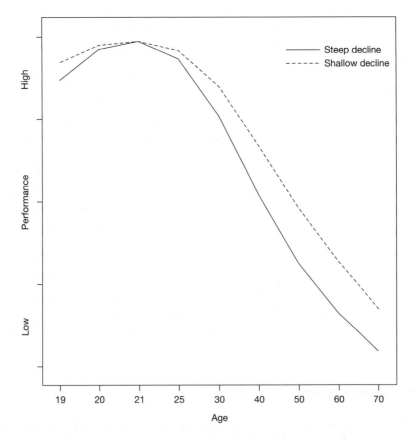

Figure 2.3 Hypothetical curves illustrating effects of bioenergic loss on events associated with shallow or steep performance declines with age

michael stones

that age at peak performance ought to be an inverse function of the steepness of performance decline at the Masters level.

The data analyzed were mean ages (pooled over sex and within event categories) of 1993 world open-class record holders in track and field and swimming. Findings for track and field showed older ages for world record holders in the sprints and racewalks than in the hurdles, jumps, and throws, with longer distance runners being of intermediate age. Findings for swimming showed record holders in the butterfly to have younger ages than in any other stroke, with increasing event distance in freestyle swimming (i.e., from 50–1500m) associated with decreasing ages of the record holders. Consequently, the findings support a continuity model because age at peak performance related inversely to established steepness of performance decline in older athletes.

Study 1

Study 1 attempted to replicate Stones and Kozma's (1996) findings with September 2008 male and female world open-class records in track and field and swimming. Predictions were the same as in the earlier study: older ages at peak performance in the sprints and racewalks than in other categories of short and long duration track and field events; younger age at peak performance in longer than shorter freestyle swims; younger age at peak performance in butterfly events than in other swimming strokes over comparable distances.

Because some athletes held multiple world records, statistical analysis was by a multilevel modeling, a procedure discussed more fully in subsequent sections of the chapter. Briefly, the models identified athletes as a random variable with multiple records by the same athlete a repeated measure. Initial models tested first-order autoregressive covariance structures for the repeated measure; however, the absence of significant evidence for autoregression led to replacement by scaled identity structures.

Fixed effect terms for the analysis of the track and field records were event categories as a factor, centered sex as a covariate, and the event categories by sex interaction. The event categories included single-step jumps, hurdles (100–400m), middle-distance metric runs (800–10,000m), sprints (110–400m), and middle-distance racewalks (3,000–20,000m). These are events with male and female records ratified by the International Amateur Athletics Federation (IAAF) for two decades or more.

The findings showed significant effects for athletes ($p<.005$) and event categories ($p<.05$). With the racewalks used as reference category (i.e., with a mean age at peak performance of 31.1 years), the relative mean ages for the other categories were:

- 5.4 years younger for the jumps ($p<.05$);
- 6.1 years younger for the hurdles ($p<.05$);
- 7.2 years younger for the middle-distance runs ($p<.005$);
- 2.9 years younger for the sprints (nonsignificant).

These findings replicate those of Stones and Kozma (1996), with younger age at peak performance in the racewalks and sprints than in the jumps, hurdles, and middle-distance runs. Although the generality of the findings does not include events with records ratified by the IAAF after 1990, the progression of records for the latter suggests a relative paucity of elite female competitors (e.g., pole vault and triple jump). With this limitation borne in mind, the findings support a continuity model suggesting that events with older ages at peak performance are those with lower performance loss with age in older athletes.

Unlike track and field, in which only a minority of world records improved in 2008, approximately 34 per cent of short-course and 63 per cent of long-course swimming records were broken during that year. Reasons for this improvement include new swimwear and a deeper-than-standard pool used in the Beijing Olympics that respectively reduce friction and turbulence. Consequently, the analyses of age at peak performance in swimming distinguish between pre-2008 and 2008 records.

The first such analysis examined age at peak performance in freestyle swimming, which is the only stroke associated with a full range of short and long distance events (i.e., 50–1,500m). Fixed effect terms included centered estimates of event distance, sex, course (short and long), date of record, and all two-way interactions between event distance and the other terms. Effects were significant for athletes ($p<.02$), distance ($p<.001$), date ($p<.001$), and distance by course ($p<.001$). The significant parameter estimates showed the following:

- Mean age was 4.4 years younger for the 1,500m swim compared with the 50m event;
- The mean age for records set in 2008 was 6.3 years younger than for records set in earlier years;
- Mean ages were younger at longer distances in long rather than short pools.

Probably the most dramatic finding is the younger mean age of holders of records set in 2008 compared with previous years. This finding suggests that technological advancement in swimwear and pool design may offset the potential to

26
michael stones

compensate for age change through practice and experience. However, the findings with respect to event distance replicate those by Stones and Kozma (1996): age at peak performance was younger in longer events. This finding adds to support for the continuity model.

The second analysis examined effects of swimming stroke on age at peak performance. Fixed effect terms included stroke as a factor; centered estimates of distance (50–200m), sex, course, and date of record as covariates; all two-way interactions between stroke and the other terms. The findings were significant for athletes ($p<.03$) and the stroke by date interaction ($p<.01$). With the butterfly as reference category, parameter estimates for the interaction showed significantly higher differences in mean ages between pre-2008 and 2008 records for the freestyle ($p<.01$) and breaststroke ($p<.005$), with the backstroke showing a similar but nonsignificant trend. These findings suggest that the ages of world-record holders in the butterfly decreased more in 2008 than did the ages of world-record holders in other strokes.

Probable reasons for the significant interaction include technological advances, referred to previously, that may affect butterfly swimmers more than other strokes. In order to illustrate age differences with records accomplished under optimal conditions, Figure 2.4 gives the mean ages of holders of records attained in 2008 in Olympic-sized pools for distances up to and including 200m. A fixed effects analysis with repeated measures showed significance for repeated records ($p<.02$) and strokes ($p<.05$), but not for sex or its interaction with strokes. The mean age of record holders in butterfly events was 2.5 years younger than in the other strokes. Consequently, the findings from this subset of world records are consistent with those of Stones and Kozma (1996).

The findings reported in this section provide support for the continuity model with respect to track and field records, distance effects in freestyle swimming, and differences across swimming strokes in 2008 records set in Olympic-sized pools. These findings replicate those with 1993 world records by showing that age at peak performance mirrors established performance on age declines at the Masters level. This mirrored inversion suggests that the effects of bioenergic processes contribute similarly to both age at peak performance and performance loss beyond that age. Although Tanaka and Seals (2008) noted that age effects on athletic performance are visible among athletes over 35 years of age, Study 1 showed such visibility in athletes as young as 20 years of age.

Discussion of implications of these findings will conclude with a comment on interpretation. Problems of interpretation with Masters-level data occur because of confounding between age and cohort. Differences between cohorts include rate of participation, and the quantity and quality of training. Record holders in open-class competition belong to a single cohort: therefore, there is no

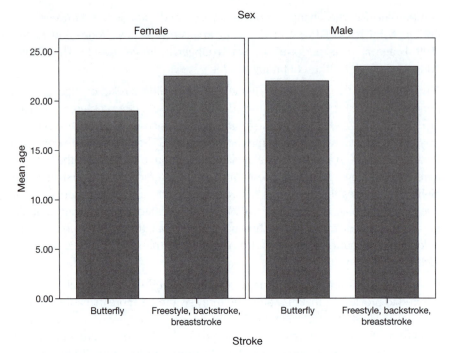

Figure 2.4 Mean ages of holders of 50–200m world records set in 2008 in Olympic-sized pools

confounding by cohort differences. The absence of such confounding makes the trends simpler to interpret as aging effects, and suggest consistency in bioenergic losses across the full spectrum of adult performance age.

STATISTICAL METHODOLOGY

Three main issues about statistical methodology relate to the choice of statistical model, adaptations to reduce error, and systemic effects. This section provides critical discussion of these issues.

Statistical models reviewed by Stones and Kozma (1986b) included exponential, second-order polynomial and proportionate change models. Recent research favors polynomial and proportionate change models (Baker et al., 2003; Fair, 2007a; Young et al., 2008). The model described by Fair (2007a) includes features of all three.

Fair (2007a) applied his model to discrete and pooled events within track and field, swimming, and chess. The unit of measurement was yearly rate of

michael stones

proportionate performance decline, which is equal to the difference between adjacent years in logarithms of performance (i.e., $P_i/P_j = \log[P_i] - \log[P_j]$). His model imposes a linear decline until a transition age, after which it becomes quadratic. Parameters of the model estimate linear rate, the transition age, and quadratic rate. Fair (2007a) noted a collinear relationship between the transition age and quadratic parameter, with lower values of the former associated with lower values of the latter. Fair's (2007a) model falls under a discontinuity rubric because it assumes differential age trends from before to after the transition age.

Adaptations made by Fair (2007a) and others to remove confounds or improve fit include the exclusion of records considered unwanted or atypical. Starkes et al. (1999) used exclusion to reduce confounds. For purposes of comparison with longitudinal records, they excluded repeat records by the same competitors from mainly cross-sectional compilations, thereby removing longitudinal elements from such data. Other researchers used exclusion to improve fit. Fairbrother (2007) removed outliers in iterative analyses of 1500m swimming records until a final model identified no residual outliers. Fair (2007a) excluded records inferior in performance to those by athletes older in years. His rationale was that any age trend other than performance decline with age must necessarily reflect sampling error.

Although exclusion of outliers to improve fit generally falls within accepted practice, modeling of elite performance presents unusual considerations. First, age is a stronger predictor of performance in elite athletes than in average competitors (Seiler et al., 1998). This higher age dependency should make it less necessary to exclude cases based on unexplained residuals. Second, the expected distribution of elite performance is not a normal bell-shaped curve but a narrow wedge at its upper extreme. Because all such performances are outliers on a normal curve, the intent of modeling elite performance is to explain data fully composed of outliers. To exclude outlying outliers might seem a bridge too far. Third, all elite performers are stars, but a few are superstars. Fair (2007a) excluded chess superstar Garry Kasparov from his analyses because his rating was such that 'no sensible line could have been fitted using this value and his age' (p. 42). Does the sacrifice of a superstar to protect the fit of a statistical model seem sensible? Would it not be sensible to reconsider the authenticity of the model?

Systemic effects include data nesting within compilations of records formerly treated as cross-sectional for purposes of analysis. Nested data include records held by the same athlete in more than one age category or event. Perusal of WMA 2007 world age-class track and field records, described in the next section, showed that nesting encompassed nearly 50 per cent of all records analyzed. Examples include a male athlete with records for the 100m at ages 50–55 and

55–60 years, and in the 200m at 55–60 years; a female athlete with records for the 3,000m run at ages 70–75 and 75–80 years, and in the 5,000m run at 75–80 years. Swimming records also show a high occurrence of nesting. FINA 2007 short-course world records show that one swimmer held 21/36 world records at the 70–74 and 75–79 year levels that encompassed all distances, three strokes, and the individual medley. Because nesting can give rise to random correlated effects, analysis that fails to encompass nesting fails to account for correlated error.

Previous studies of athletic records either ignored nesting or excluded repeat records from the data analyzed (e.g., Young et al., 2008). Although the development of analytic methods that account for nesting came too late for the early researchers, they have been available for some years (e.g., SPSS introduced mixed linear analysis in 2002). This form of analysis should be a requirement with nested data. Study 2 illustrates such modeling by separating cross-sectional from longitudinal trend in analysis of world age-class track and field records.

STUDY 2: MULTILEVEL ANALYSIS OF TRACK AND FIELD RECORDS

The intent of Study 2 was to compare age by event interactions within a compilation of athletic records that encompasses cross-sectional and longitudinal trends. The data were WMA 2007 age-class world records by men and women. The criteria for selection of events were as follows: records by both sexes for age classes 35–39 to 85–89 years; identical events for both sexes and across all age levels; comparable distances between middle distance running and racewalking events. Although records for the hurdles, pole vault, and throws failed to meet all the criteria, the following events were included in analysis:

- Jumps: long jump, high jump, triple jump;
- Sprints: 100m, 200m, 400m;
- Runs: 3,000m, 5,000m, 10,000m;
- Racewalks: 3,000m, 5,000m, 10,000m.

Inspection of the data showed that 51.1 per cent of records were by discrete athletes and 49.9 per cent by athletes with repeated records. Athletes with repeated records had a mean of 2.37 records (s.d. = 1.34). Cross tabulations with the presence or absence of nesting within athletes showed no differences across sex and age categories but a higher level of nesting in the racewalks than in other event categories ($p < .02$).

30

michael stones

The athletic performance measure used in analysis was a centered logarithmic scale based on log[Distance] for the jumps and log[1/Time] for the sprints, runs, and racewalks. Higher scores on the measure signify higher levels of performance irrespective of event. Centering the measure on mean levels for discrete events and sex rendered performance means of zero but with no effect on dispersion within events for either sex.

The multilevel model tested against the performance measure included random, repeated, and fixed effects. The random intercept comprised athletes; records nested within athletes were a repeated effect. Fixed effects included event categories along with covariates that included sex and age terms representing across-athlete and within-athlete effects. Across-athlete age was the mean age over all records held by an athlete. Within-athlete age was the age associated with a discrete record and centered on the athlete's mean age. These two terms respectively represent cross-sectional and longitudinal trends. The effects tested in the model were all the main effects and first-order interactions.

Data input modifications to facilitate interpretation of the intercept, main effects, and interactions were consistent with previous literature (e.g., Gelman & Hill, 2007, pp. 55–57). First, all the fixed terms were centered on zero to ensure a null intercept in that component of the model. Second, values of the fixed terms were reduced to unitary differences between adjacent intervals in order to facilitate comparison of parameter estimates. Consequently, across and within-athlete ages were modified such that one unit represented a five-year interval; likewise, sex categories differed by one unit.

Initial analyses tested the covariance structure within the repeated effect. Because an autoregressive structure yielded nonsignificant random parameters, it was replaced by scaled identity that assumes no such relationship. The overall results with this structure showed good fit to the data, with predicted scores that correlated with the performance measure at .98. Parameters for both the random intercept (i.e., athletes) and repeated measure terms were significant ($p<.001$), with fixed effects significant for (1) across-athlete age ($p<.001$), (2) within-athlete age ($p<.001$), (3) across-athlete age by within-athlete age ($p<.001$), (4) across-athlete age by sex ($p<.001$), and (5) across-athlete age by event categories ($p<.024$).

The significant parameter estimates for across-athlete age (-.032) and within-athlete age (-.029) were of similar magnitude, indicating comparable overall trends between cross-sectional and longitudinal performance loss with age. However, the significant parameter estimate for their interaction (-.005) suggests a higher performance loss for within-athletes at older than younger mean ages. Figure 2.5 illustrates this interaction wherein longitudinal performance loss is higher in older than younger athletes.

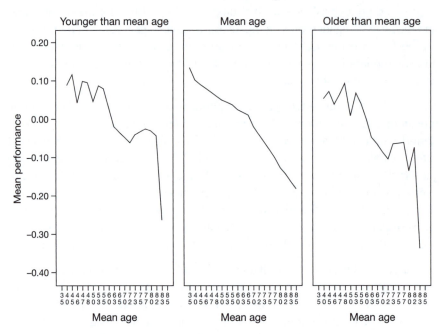

Figure 2.5 Mean centered log[performance] by across-athlete and within-athlete age

The significant parameter estimate for the across-athlete age by sex interaction (-.009) indicates that females show greater performance loss than males at higher cross-sectional ages. Figure 2.6 shows this trend. The nonsignificant within-athlete age by sex interaction provides no evidence for sex differences in longitudinal age trend.

With jumps as the reference category, parameter estimates for the across-athlete age by event categories interaction were significant for the racewalks (.005) and sprints (.004) but not the runs (.001). Figure 2.7 shows these cross-sectional trends wherein performance loss with age is lower in the racewalks and sprints than in the runs and jumps. The absence of a significant interaction between within-athlete age event and event categories provides no evidence of a comparable interaction with longitudinal trend.

The cross-sectional findings of lower performance loss in the racewalks and sprints than in the longer runs and jumps, and in males than females, echo those from studies dating back to 1980. Although the present findings showed no comparable interactions with longitudinal data, this was likely due to the relative

michael stones

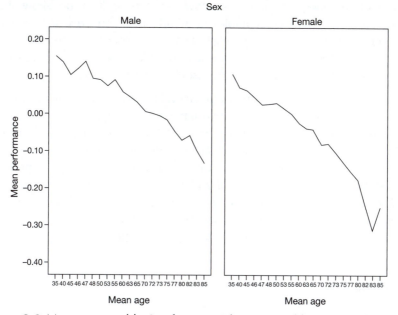

Figure 2.6 Mean centered log[performance] by across-athlete age and sex

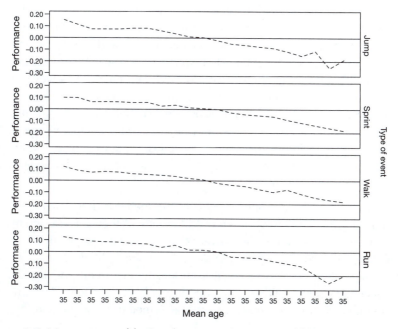

Figure 2.7 Mean centered log[performance] by across-athlete age and event categories

statistical modeling of age trends

brevity of the longitudinal spans. However, the findings do show mean performance losses of comparable magnitude between cross-sectional and longitudinal trend, with higher longitudinal loss associated with older cross-sectional age. Consequently, late life loss in athletic performance appears to reflect age change more than factors associated with participation.

Study 2 was probably the first to use multilevel modeling to examine correlated error in athletic records nested within athletes. Although preliminary findings failed to detect autocorrelation, parameter estimates associated with athletes and repetitions of records indicate that both are significant sources of random variation that affect independence within the data array. What is probably most striking, however, is the facility of multilevel modeling to partition the findings into those associated with cross-sectional and longitudinal trend. Such separation might now seem not only sensible but indispensable for future research on age trends in sports that include nesting within athletes.

CONCLUSIONS

This chapter began with reminiscence and will end with reflection. A good question to reflect upon is the following: What do we know about statistical modeling of records by Masters Athletes that was unknown by the end of the first decade of research in 1985? The findings summarized in section one show that the primary age trends, first authenticated in track and field and swimming, endure today. These sports continue to be popular for research purposes because of the quality and comprehensiveness of age-class records. However, other research extended the range of sports examined. Seiler et al. (1998) studied indoor rowing, Sowell and Mounts (2005) researched triathlon athletes, Fair (2007b) analyzed baseball records, and Stones (2001) compared age ranges for elite competitors in multiple sports, to cite just some examples. This chapter certainly would have gained in coverage had its author decided to forgo depth of discussion for greater breadth.

If the quality of findings is only as good as the methods used for discovery, it is encouraging that past discoveries proved so robust. Although regression models have progressed over time, the arrival of mixed linear analysis threatens the viability of earlier forms with nested data. As shown in Study 2, such analysis enables separation of cross-sectional from longitudinal trend. Implications are that researchers should neither ignore nesting nor have need to nullify correlated error through exclusion of repeat records, as sometimes was done in the past.

Consensus about the robustness of age trends in Masters Athletes is less evident in interpretations of those trends. Stones and Kozma (1986a) proposed a

michael stones

bioenergic model; Tanaka and Seals (2008) emphasized physiological changes; Seiler et al. (1998) argued physics over physiology; Weir et al. (2002) cite differences in training. Study 1 suggests a different perspective based on findings that age at peak performance proved to be continuous with expected performance loss among Masters Athletes.

This perspective proposes continuity in transition throughout adulthood. Such transition can be symmetrical or asymmetrical, with compensation through practice and experience balanced against aging effects that affect some sports activities more than others. As viewed through a continuity lens, transitions from open-class athlete to Masters Athlete, and younger to older Masters Athlete are conventional rather than natural. Surely it is more sensible to study adulthood as a continuum rather than in artificial stages. It would be ironic if the study of age at peak performance helped to clarify interpretations of age changes in Masters Athletes; from a continuity perspective, that could happen.

REFERENCES

Astrand, P.-O., & Rodahl, K. (1977). *Textbook of work physiology: Physiological bases of behavior*. New York: McGraw-Hill (2nd edition).

Baker A.B., Tang, Y.Q., & Turner, F.M. (2003). Percentage decline in masters superathlete track and field performance with aging. *Experimental Aging Research*, 2003, 29, 47–65.

Cooper, K.H. (1977). *The aerobics way*. New York: M. Evans & Co.

Donato, A.J., Tench, K., Glueck, D.H., Seals, D.R., Eskurza, I., & Tanaka, H. (2003). Age and gender interactions in physiological functional capacity: Insight from swimming performance. *Journal of Applied Physiology*, 94, 764–769.

Fair, R. (2007a). Estimated age effects in athletic events and chess. *Experimental Aging Research*, 33, 35–57.

Fair, R. (2007b). Estimated age effects in baseball. *Cowles Foundation Discussion Paper # 1536*. Retrieved from http://cowles.econ.yale.edu/.

Fairbrother, J.T. (2007). Prediction of 1500-m swimming times for older masters all-American swimmers. *Experimental Aging Research*, 33, 461–471.

Gelman, A., & Hill, J. (2007). *Data analysis using regression and multilevel/hierarchical models*. N.Y.: Cambridge University Press.

Hartley, A., & Hartley, J.T. (1984a). Performance changes in champion swimmers aged 30–70 years. *Experimental Aging Research*, 10, 141–147.

Hartley, A., & Hartley, J.T. (1984b). In response to Stones and Kozma: Absolute and relative decline with age in champion swimming performances. *Experimental Aging Research*, 10, 151–153.

Henry, F.M. (1955). Prediction of world records in running sixty yards to twenty-six miles. *Research Quarterly*, 26, 147–156.

Holmér, I. (1974). Energy cost of arm stroke, leg kick, and the whole stroke in competitive swimming. *European Journal of Applied Physiology*, 33, 105.

Jokl, P., Sethi, P.M., & Cooper, A.J. (2004). Master's performance in the New York City Marathon 1983–1999. *Sports Medicine*, 35, 1017–1024.

Magel, J.R. (1970). Propelling force measured during tethered swimming in the four competitive swimming styles. *Research Quarterly*, 41, 68–74.

Moore, D.H. (1975). A study of age group track and field records to relate age to running speed. *Nature*, 253, 264–265.

Mundle, P., & Brieger, K. (1979). Masters records. *Runner's World*, 14, 88–93.

Rowe, J.W., & Kahn, R.L. (1987). Human aging: Usual and successful. *Science*, 237, 143–149.

Rybash, J.M., Roodin, P.A., & Santrock, J.W. (1991). *Adult development and aging* (2nd edition). Dubuque, IA: Wm. C. Brown.

Salthouse, T.A. (1976). Speed and age: Multiple rates of age decline. *Experimental Aging Research*, 2, 349–359.

Seiler, K.S., Spirduso, W., & Martin, J.C. (1998). Gender differences in rowing performance and power with aging. *Medicine and Science in Sports*, 30, 121–127.

Shephard, R.J. (1978). *Physical activity and aging*. Chicago: Yearbook Medical Publishers.

Smith, E.L., & Gilligan, G. (1983). Physical activity prescription for the older adult. *The Physician and Sportsmedicine*, 11, 91–101.

Spirduso, W.W. (1980). Physical fitness, aging, and psychomotor speed. *Journal of Gerontology*, 35, 850–865.

Starkes, J.L., Weir, P.L., Singh, P., Hodges, N.J., & Kerr, T. (1999). Aging and the retention of sports expertise. *International Journal of Sports Psychology*, 30, 283–301.

Stones, M.J., (2001). Age differences in sports performance. In N.J. Smelser & P.B. Baltes (Eds.), *International Encyclopedia of the Social and Behavioral Sciences*. Amsterdam: Elsevier.

Stones, M.J., & Kozma, A. (1980). Adult age trends in record running performances. *Experimental Aging Research*, 6, 407–416.

Stones, M.J., & Kozma, A. (1981). Adult age trends in athletic performances. *Experimental Aging Research*, 7, 269–279.

Stones, M.J., & Kozma, A. (1982a) Cross-sectional, longitudinal, and secular age trends in athletic performances. *Experimental Aging Research*, 8, 185–188.

Stones, M.J., & Kozma, A. (1982b). Sex differences in changes with age in record running performances. *Canadian Journal on Aging*, 1, 12–16.

Stones, M.J., & Kozma, A. (1984a) Longitudinal trends in track and field performances. *Experimental Aging Research*, 10, 107–110.

Stones, M.J., & Kozma, A. (1984b). In response to Hartley and Hartley: Cross-sectional age trends in swimming records: decline is greater at the longer distances. *Experimental Aging Research*, 10, 149–150.

Stones, M.J., & Kozma, A. (1986a). Age trends in maximal physical performance: Comparison and evaluation of models. *Experimental Aging Research*, 12, 207–215.

Stones, M.J., & Kozma, A. (1986b). Age by distance effects in running and swimming records: A note on methodology. *Experimental Aging Research*, 12, 203–206.

Stones, M.J., & Kozma, A. (1996). Activity, exercise and behavior. In J. Birren & K.W. Schaie (Eds.), *Handbook of the psychology of aging* (5th ed.). Orlando, FL.: Academic Press.

Sowell, C.B., & Mounts, W.S. (2005). Age, ability and performance: Conclusions from the world triathlon Ironman championship. *Journal of Sports Economics*, 6, 78–97.

Tanaka, H., & Seals, D.R. (1997). Age and gender interactions in physiological functional capacity: Insight from swimming performance. *Journal of Applied Physiology*, 82, 846–851.

Tanaka, H., & Seals, D.R. (2003). Invited review: Dynamic exercise performance in Masters athletes: insight into the effects of primary human aging on physiological functional capacity. *Journal of Applied Physiology*, 95, 2152–2162.

Tanaka H., & Seals, D.R. (2008). Endurance exercise performance in Masters athletes: age-associated changes and underlying physiological mechanisms. *Journal of Physiology-London*, 586, 55–63.

Weir, P.L., Kerr, T., Hodges, N.J., McKay, S.M., & Starkes, J.L. (2002). Masters swimmers: How are they different from younger elite swimmers. An examination of practice and performance patterns. *Journal of Aging and Physical Activity*, 10, 41–63.

Young, B.W., & Starkes, J.L. (2005). Career-span analyses of track performance: Longitudinal data provide a more optimistic view of age-related performance decline. *Experimental Aging Research*, 31, 69–90.

Young, B.W., Weir, P.L., Starkes, J.L., & Medic, N. (2008). Does lifelong training temper age-related decline in sport performance. Interpreting differences between cross-sectional and longitudinal data. *Experimental Aging Research*, 34, 27–48.

SECTION TWO

AGING, PERFORMANCE, AND THE ROLE OF CONTINUED INVOLVEMENT

CHAPTER THREE

PEAK EXERCISE PERFORMANCE, MUSCLE STRENGTH, AND POWER IN MASTERS ATHLETES

HIROFUMI TANAKA

The age profile in the United States is shifting very rapidly as the first baby boomers move toward an older age structure crossing the retirement age threshold. It is estimated that by year 2030, one in every five Americans will be 65 years and older. Aging is associated with declines in functional capacity and increased risks of developing chronic diseases. However, being old is not the inevitable state of functional disability and illness. In fact, functional and health-related changes that we often associate with aging are in large part due to physical inactivity (Skinner et al., 1982). Short-term inactivity through bed rest and weightlessness produces substantial loss of muscle mass and strength (Kortebein et al., 2007; Volpi et al., 2004), whereas progressive strength training induces muscle hypertrophy and increases muscle strength and power (Anton et al., 2006; Miyachi et al., 2004). In this context, Masters Athletes are an effective experimental model of 'primary aging' and have challenged the negative stereotype of aging (Tanaka & Higuchi, 1998; Tanaka & Seals, 2003; see also Horton, Chapter 8). In essence, Masters Athletes represent the other extreme end of an aging distribution, a complete opposite to the frail elderly. This review will focus on the age-associated changes in muscle strength and power in Masters Athletes.

EXPERIMENTAL STUDIES ON MASTERS ATHLETES

Obviously in the data of athletic records we have a store of information available for physiological study. Apart from its usefulness, however, I would urge that the study is amusing. Most people are interested, at any rate in England and America, in some type of sport. If they can be made to find it more interesting, as I have found it, by a scientific contemplation of the

41

things which every sportsman knows, then that extra interest is its own defense.

(Hill, 1925)

The Nobel laureate, A.V. Hill, pioneered the use of athletic performance data to examine the relation between maximal speed and racing distance (Hill, 1925). This experimental approach was later adapted for the study of aging, originally by Lehmann, who examined the relation between age and peak performance (Lehmann, 1953). Since then, a number of investigators have utilized this particular approach to address a number of questions pertinent to aging (Schulz & Curnow, 1988; Stones & Kozma, 1981; Tanaka & Seals, 1997; Tanaka & Seals, 2003). The increasing popularity of the use of athletic performance data to examine physiological aging can be attributed to a number of experimental advantages that the study of Masters Athletes could provide as shown below.

1) Extrinsic factors (e.g., deconditioning, chronic degenerative diseases) that often confound the intrinsic aging process can be minimized in Masters Athletes.
2) Athletic performance data contain rich sources of physiological information allowing insight into age-related changes in physiological functional capacity.
3) Masters Athletes maintain high levels of motivation and the drive to succeed.
4) Athletic performance data are very reliable as they are collected in well-controlled environments due to elaborate rules and closely monitored competitions (Henry, 1955).
5) Masters Athletes provide the gauge at which we can assess the physiological ceiling at older ages.
6) It is simply interesting and amusing. As illustrated by the popularity of the Olympic Games, people are fascinated by the athletic achievements and upper limits of what athletes can do in athletic events.

As in any other research field, experimental studies utilizing Masters Athletes and athletic performance data are not without limitations/weaknesses. Some of the major limitations associated with this experimental approach are: (a) secular changes related to more rapid growth of older Masters Athletes (Jokl et al., 2004); (b) influence of sociocultural factors (e.g., women were not allowed to compete in many athletic events in the past); (c) potential problem of generalizability to the entire aging population. Masters Athletes, at least those in the US, are mostly white and well educated (Wright & Perricelli, 2008), and may not be a true representative sample; (d) influence of non-physiological factors (e.g., changes and improvements in equipment and techniques) acting on athletic perform-ance; and (e) potential fundamental differences between lifelong Masters

Athletes and newcomers (i.e., possible differences in the effects of long-term and short-term training).

PHYSIOLOGICAL FUNCTIONAL CAPACITY WITH AGING

Physiological functional capacity (PFC) can be defined as the ability to perform the tasks of daily life and the ease with which these tasks can be performed (Tanaka & Seals, 2003). Determination of the effects of biological aging on physiological functional capacity is difficult because of the confounding factors that often change concomitant with aging (e.g., decline in physical activity, increase in chronic degenerative disease). Analysis of changes in peak sport performance is an effective approach to assess how physiological functional capacity is affected by the aging process as changes observed with advanced age in these Masters Athletes are thought to reflect mainly the results of primary aging. We have previously reported how physiological functional capacity, as assessed by running and swimming endurance performance, declines with advancing age (Tanaka & Higuchi, 1998; Tanaka & Seals, 1997; Tanaka & Seals, 2008; see also Stones, Chapter 2). Another, and arguably more important, component of physiological functional capacity in relation to aging is muscular strength and power. The age-associated decline in peak muscular power has important clinical and functional implications for the elderly. The ability to perform many activities of daily living may be compromised by low muscle strength and power even in healthy elderly persons. Because we cannot normally change the physical demands of our daily work with aging, a reduction in PFC means that aging workers labor closer to their maximal capacity, and that could result in chronic fatigue and other health problems (WHO Study Group, 1993). Additionally, a reduction in PFC in aging workers has economical implications. There was an interesting study published in the 1960s describing the trend whereby the average earnings of forest workers dropped progressively with advancing age despite the fact that average daily work time remained unchanged (Kilander, 1962).

AGE-RELATED DECLINES IN MUSCLE STRENGTH AND POWER IN SEDENTARY ADULTS

Skeletal muscle strength, one of the representative measures of functional capacity, begins to decline after age 30, with a more exponential decrease in strength after the age of 50 (Grimby & Saltin, 1983; Quetelet, 1842). Between the ages of 30 and 80, humans lose an average of 30–40 per cent of their muscle strength (around 40 per cent in the leg and back muscles and 30 per cent in the arm muscles; Grimby & Saltin, 1983; Holloszy & Kohrt, 1995). The primary mechanisms underlying this decrease in muscle strength with age, which is commonly referred to as 'sarcopenia', is a decline in muscle mass, as well as a decrease in muscle strength per unit muscle cross-sectional area (i.e., neural activation or muscle quality; Dutta & Hadley, 1995; Volpi et al., 2004). The consequences of sarcopenia can be extensive because there is an increased susceptibility to falls and fractures, impairment in the ability to thermoregulate, a decrease in basal metabolic rate, as well as an overall loss in the functional ability to perform daily tasks (Dutta & Hadley, 1995).

AGE-RELATED DECLINES IN MUSCLE STRENGTH AND POWER IN ENDURANCE-TRAINED MASTERS ATHLETES

It has been well established that regular aerobic exercise improves a number of cardiovascular functions, risk factors for cardiovascular disease, and overall functional capacity in older adults (Holloszy & Kohrt, 1995; Mazzeo & Tanaka, 2001). Because of the major role that habitual exercise plays in determining physical function, it is reasonable to speculate that regular aerobic exercise could have beneficial effects on muscle mass and muscle strength. However, as is apparent in the stereotypic appearance of Masters endurance-trained runners, regular aerobic exercise does not induce obvious muscle hypertrophy (Sugawara et al., 2002; Volpi et al., 2004) although an increase in muscle fiber area has been observed after intense endurance training in older adults (Coggan et al., 1992). Moreover, endurance training does not appear to attenuate or prevent loss of muscle mass with increasing age (Sugawara et al., 2002). Furthermore, rates of age-related decreases in anaerobic power, as assessed by vertical jumping performance, are essentially the same between sedentary and Masters Athletes (Grassi et al., 1991). Although a lack of effects of endurance training in modulating age-associated loss of muscle mass is certainly disappointing, it is important to recognize that muscle mass is not the only determinant of muscle function. It is possible that regular aerobic exercise induces beneficial effects on 'muscle quality', including neuromuscular components, in older adults (Volpi et al., 2004).

AGE-RELATED DECLINES IN MUSCLE STRENGTH AND POWER IN STRENGTH-TRAINED MASTERS ATHLETES

The age at which peak performance in lifting events is achieved occurs between 28 and 31 years of age (Schulz & Curnow, 1988). After that age, lifting performance starts to decline with advancing age. What is the temporal pattern of the decline in peak dynamic muscular power with age in strength-trained Masters Athletes? Is it a continuous (linear) decline or a curvilinear decrease with aging? In order to address these questions, we performed retrospective analyses of the data compiled from US weightlifting and powerlifting records (Anton et al., 2004). Weightlifting consists of two main events. The snatch is performed in a continuous movement from the bar on the floor to the fully extended arm position above the head, whereas the clean & jerk involves lifting from the platform to the shoulders in one motion, then thrusting the bar into a position overhead, and finally bringing feet together to complete the lift. Powerlifting consists of three events: (a) deadlift, (b) squat, and (c) bench press. In the deadlift, a competitor lifts a barbell off the ground from a bent-over position until the torso is fully upright. The squat involves lowering the torso by bending the knees and hips until the hip joint comes lower than the knee joint, and then standing back up. In the bench press, while lying on his or her back, a competitor lowers a barbell to the chest and then pushes it back up until the arm is fully extended.

Weightlifting performance declined curvilinearly with advancing age in both men and women, whereas age-related decrease in powerlifting performance was linear (see Figure 3.1). Additionally, the magnitude of age-related declines in weight-lifting performance was substantially greater than in powerlifting. Age-related declines in physiological functional capacity can be attributed to overall decreases in a number of physiological functions. Each lifting event is unique in that the degree to which each of the physiological systems is involved differs considerably. Weightlifting events require quickness and explosive power as well as more complex and exquisite neuromuscular coordination to lift the load. It is also critical to possess excellent balance throughout the lift. In contrast, speed is not a critical factor for powerlifting events, and the movement required in each event is relatively simple. These results suggest that more complex and explosive tasks that require a greater involvement of various physiological functions demonstrate greater decreases in physiological functional capacity with advancing age (Anton et al., 2004).

Another interesting observation is that age-associated reductions in performance were greater in women than in men in weightlifting events, whereas the rate and magnitude of age-related decrease in powerlifting performance were not different between the genders. This observation is consistent with physiological research indicating that women may undergo greater age-related reductions in muscle fiber type, shortening velocity at the single fiber level (Krivickas et al., 2001). Certainly, it is possible that sociocultural factors contributed to these observations, because the explosive nature of weightlifting may have discouraged more women from competing in these events.

45

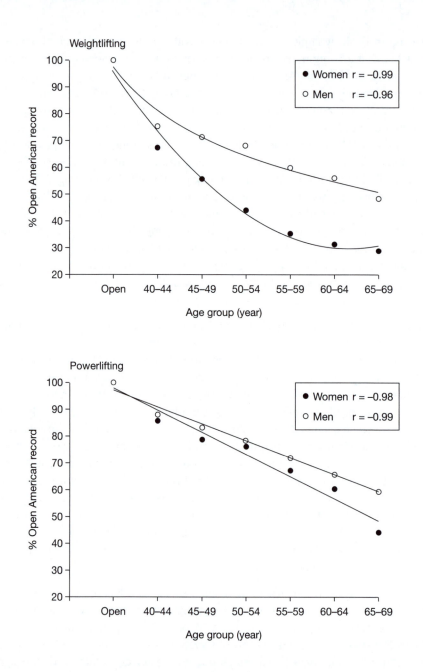

Figure 3.1 Age-related decreases in weightlifting (an average of snatch and clean & jerk) and powerlifting (an average of deadlift, squat, and bench press) performance records in men and women (Anton et al. 2004) (reproduced with permission from Lippincott, Williams & Wilkins)

hirofumi tanaka

SEDENTARY VS. STRENGTH-TRAINED MASTERS ATHLETES

A clinically and functionally important question is whether the rate of decline in muscle strength and power with age is attenuated or absent in adults who perform regular resistance exercise. The notion that strength training performed on a daily basis will attenuate or prevent loss of muscle strength with age is a very positive message from the public health standpoint, and such notions have been promoted and described in textbooks (Wilmore & Costill, 2004). Surprisingly, only a few published studies are available to provide insight into this issue. In a study that compared Masters weightlifters and healthy untrained adults varying widely in age (Pearson et al., 2002), both peak muscle isometric strength and peak lower-limb explosive power declined with increasing age at a similar relative (per cent) rate in the weightlifters and sedentary controls. When the data in peak muscle power was expressed in absolute unit (in W/year), the rate of decrease was ~60 per cent greater in strength-trained adults (Pearson et al., 2002). Similar relative rates of age-related decline in anaerobic power have been reported between power-trained Masters Athletes and sedentary peers (Grassi et al., 1991). Thus, the available evidence is not consistent with the notion that regular strength training would prevent loss of muscle strength and power with increasing age. However, it is important to note from the standpoint of preventive gerontology that the absolute levels of muscle strength and power in strength-trained adults are substantially higher than those of their sedentary peers throughout the adult age range (Pearson et al., 2002). Accordingly, strength-trained adults possess higher levels of physiological functional capacity and lower risks of premature morbidity than sedentary adults at any age.

Interestingly, these results obtained in muscle strength and power in relation to age are very similar to the trends observed in another important index of physiological functional capacity: maximal aerobic capacity (see Hawkins, Chapter 4). Early investigations reported that the rate of decline in maximal aerobic capacity with age is substantially smaller in endurance-trained adults than in sedentary adults (Heath et al., 1981; Kasch et al., 1990). In marked contrast to these early observations, a series of more recent investigations using a variety of experimental approaches (i.e., meta-analyses, laboratory-based cross-sectional, and longitudinal study design) revealed that the absolute (i.e., ml/kg/min/year) rate of decline in VO_2max with increasing age was *greatest* in endurance-trained adults, next greatest in active adults, and lowest in sedentary adults (FitzGerald et al., 1997; Pimentel et al., 2003; Tanaka et al., 1997). When expressed as per cent or relative decrease from mean levels at age ~25 years, however, the rate of decline in VO_2max was similar in the three groups

47

(FitzGerald et al., 1997; Pimentel et al., 2003; Tanaka et al., 1997). Greater rates of decline in maximal aerobic capacity in endurance-trained vs. sedentary adults are presumably a result of greater baseline levels of maximal oxygen consumption as young adults and greater reductions in habitual exercise training with aging compared with sedentary adults (Eskurza et al., 2002; FitzGerald et al., 1997). Currently, we have no information regarding how strength training characteristics change with advancing age in Masters Athletes.

FUTURE RESEARCH DIRECTIONS

With the preceding analyses of the existing knowledge base in the field, we deem that a more concerted effort should be initiated and conducted by investigators in the future to:

1) Conduct more longitudinal studies of strength-trained Masters Athletes followed for many years. Currently, very few studies have addressed the longitudinal age-related changes in muscle strength performance in Masters Athletes (Meltzer, 1994). Considering the well-known limitations of the cross-sectional study design, well-designed longitudinal studies may provide different results.

2) Quantify age-related changes in strength training characteristics (mode, intensity, sets, frequency) in Masters Athletes. Masters (middle-aged) endurance athletes appear to be able to maintain vigorous exercise training for up to ten years (Pollock et al., 1987). However, there is no evidence that exercise training intensity and volume can be maintained for longer periods, especially at older ages (Pollock et al., 1987). It is not known how much, if any, strength training stimulus changes with age and whether strength-trained Masters Athletes can keep training at older ages for longer than ten years.

3) Determine the extent by which changes in strength training characteristics influence PFC. Factors that contribute to reductions in muscle strength and power with age in Masters strength athletes are incompletely understood. Although there is no question that the seemingly inevitable consequence of an overall reduction in the exercise training stimulus with age contributes to such decline, the magnitude of this influence on PFC has not been quantified (Weir et al., 2002).

4) Examine influence of sociocultural factors in determining age-related reductions in PFC (e.g., more men preferring to perform strength training than women).

5) Obtain more athletic performance data in Masters Athletes in the 'oldest-old' age range (aged >85 years), particularly centenarians. This age group

48

is the fastest growing segment of the aging population, and an increasing number of athletes are competing in this age group. Impressive athletic performance of oldest-old Masters Athletes has been reported in the running events. For example, in 2003, Fauja Singh of Great Britain set a new age-group record in marathon of 5 hours 40 minutes at the age of 92 years. In 2004, Philip Rabinowitz of Russia became the fastest centenarian by running 100m at 30.86 seconds. Equivalent athletic performance data of oldest-old strength-trained athletes are currently absent, and it is not clear what the upper limit of athletic performance is for the athletes competing in this age group.

6) Explore in greater detail the physiological mechanisms (e.g., muscle cross-sectional area, muscle fiber type changes, neuromuscular function, etc.) underlying age-related declines in strength-related PFC.

CONCLUDING REMARKS

Muscular strength and power are important components of physiological functional capacity in relation to aging. The available research indicates that endurance training does not modulate age-associated loss of muscle mass and does not support the notion that regular strength training would prevent loss of muscle strength and power with increasing age. However, it is important to note from the standpoint of preventive gerontology that the absolute levels of muscle strength and power in strength-trained adults are substantially higher than those of their sedentary peers throughout the adult age range.

REFERENCES

Anton, M.M., Cortez-Cooper, M.Y., DeVan, A.E., Neidre, D.B., Cook, J.N., & Tanaka, H. (2006). Resistance training increases basal limb blood flow and vascular conductance in aging humans. *Journal of Applied Physiology*, 101, 1351–1355.

Anton, M.M., Spirduso, W.W., & Tanaka, H. (2004). Age-related declines in anaerobic muscular performance: weightlifting and powerlifting. *Medicine and Science in Sports and Exercise*, 36, 143–147.

Coggan, A.R., Spina, R.J., King, D.S., Rogers, M.A., Brown, M., Nemeth, P.M., & Holloszy, J.O. (1992). Skeletal muscle adaptations to endurance training in 60- to 70-yr-old men and women. *Journal of Applied Physiology*, 72, 1780–1786.

Dutta, C., & Hadley, E.C. (1995). The significance of sarcopenia in old age. *Journal of Gerontology*, 50A (Special), 1–4.

Eskurza, I., Donato, A.J., Moreau, K.L., Seals, D.R., & Tanaka, H. (2002). Changes in maximal aerobic capacity with age in endurance-trained women: 7-year follow-up. *Journal of Applied Physiology*, 92, 2303–2308.

FitzGerald, M.D., Tanaka, H., Tran, Z.V., & Seals, D.R. (1997). Age-related decline in maximal aerobic capacity in regularly exercising vs sedentary females: A meta-analysis. *Journal of Applied Physiology*, 83, 160–165.

Grassi, B., Cerretelli, P., Narici, M.V., & Marconi, C. (1991). Peak anaerobic power in master athletes. *European Journal of Applied Physiology*, 62, 394–399.

Grimby, G., & Saltin, B. (1983). The ageing muscle. *Clinical Physiology*, 3, 209–218.

Heath, G.W., Hagberg, J.M., Ehsani, A.A., & Holloszy, J.O. (1981). A physiological comparison of young and older endurance athletes. *Journal of Applied Physiology*, 51, 634–640.

Henry, F.M. (1955). Prediction of world records in running sixty yards to twenty-six miles. *Research Quarterly*, 26, 147–158.

Hill, A.V. (1925). The physiological basis of athletic records. *Science Monthly*, 21, 409–428.

Holloszy, J.O., & Kohrt, W.M. (1995). Exercise. In E. J. Masoro (Ed.), *Handbook of Physiology*, Section 11: Aging (pp. 633–666). New York: Oxford University Press.

Jokl, P., Sethi, P.M., & Cooper, A.J. (2004). Master's performance in the New York City Marathon 1983–1999. *British Journal of Sports Medicine*, 38, 408–412.

Kasch, F.W., Boyer, J.L., Camp, S.P.V., Verity, L.S., & Wallace, J.P. (1990). The effect of physical activity and inactivity on aerobic power in older men (a longitudinal study). *Physician and Sportsmedicine*, 18, 73–83.

Kilander, K. (1962). An example of earnings in various age classes of manual labour. *Ergonomics*, 5, 291–292.

Kortebein, P., Ferrando, A., Lombeida, J., Wolfe, R., & Evans, W.J. (2007). Effect of 10 days of bed rest on skeletal muscle in healthy older adults. *JAMA*, 297, 1772–1774.

Krivickas, L.S., Suh, D., Wilkins, J., Hughes, V.A., Roubenoff, R., & Frontera, W.R. (2001). Age- and gender-related differences in maximum shortening velocity of skeletal muscle fibers. *American Journal of Physical Medicine and Rehabilitation*, 80, 447–455.

Lehmann, H. C. (1953). *Age and achievement*. Princeton, NJ: Princeton University Press.

Mazzeo, R.S., & Tanaka, H. (2001). Exercise prescription for the elderly: current recommendations. *Sports Medicine*, 31, 809–818.

Meltzer, D.E. (1994). Age dependence of Olympic weightlifting ability. *Medicine and Science in Sports and Exercise*, 26, 1053–1067.

Miyachi, M., Kawano, H., Sugawara, J., Takahashi, K., Hayashi, K., Yamazaki, K., Tabata, I., & Tanaka, H. (2004). Unfavorable effects of resistance training on central arterial compliance: a randomized intervention study. *Circulation*, 110, 2858–2863.

Pearson, S.J., Young, A., Macaluso, A., Devito, G., Nimmo, M.A., Cobbold, M., & Harridge, S.D.R. (2002). Muscle function in elite master weightlifters. *Medicine and Science in Sports and Exercise*, 34, 1199–1206.

Pimentel, A.E., Gentile, C.L., Tanaka, H., Seals, D.R., & Gates, P.E. (2003). Greater rate of decline in maximal aerobic capacity with age in endurance-trained vs. sedentary men. *Journal of Applied Physiology*, 94, 2406–2413.

Pollock, M.L., Foster, C., Knapp, D., Rod, J.L., & Schmidt, D.H. (1987). Effect of age and training on aerobic capacity and body composition of master athletes. *Journal of Applied Physiology*, 62, 725–731.

Quetelet, M.A. (1842). *A treatise on men*. Edinburgh: William and Robert Chambers.

Schulz, R., & Curnow, C. (1988). Peak performance and age among superathletes: track and field, swimming, baseball, tennis, and golf. *Journal of Gerontology*, 43, 113–120.

Skinner, J.S., Tipton, C.M., & Vailas, A.C. (1982). Exercise, physical training, and the ageing process. In A. Viidik (Ed.), *Lectures on Gerontology* (pp. 407–439). London: Academic Press.

Stones, M.J., & Kozma, A. (1981). Adult age trends in athletic performance. *Experimental Aging Research*, 7, 269–280.

Sugawara, J., Miyachi, M., Moreau, K.L., Dinenno, F.A., DeSouza, C.A., & Tanaka, H. (2002). Age-related reductions in appendicular skeletal muscle mass: association with habitual aerobic exercise status. *Clinical Physiology and Functional Imaging*, 22, 169–172.

Tanaka, H., DeSouza, C.A., Jones, P.P., Stevenson, E.T., Davy, K.P., & Seals, D.R. (1997). Greater rate of decline in maximal aerobic capacity with age in physically active vs. sedentary healthy women. *Journal of Applied Physiology*, 83, 1947–1953.

Tanaka, H., & Higuchi, M. (1998). Age, exercise performance, and physiological functional capacities. *Advances in Exercise and Sports Physiology*, 4, 51–56.

Tanaka, H., & Seals, D.R. (1997). Age and gender interactions in physiological functional capacity: insight from swimming performance. *Journal of Applied Physiology*, 82, 846–851.

Tanaka, H., & Seals, D.R. (2003). Dynamic exercise performance in Masters athletes: insight into the effects of primary human aging on physiological functional capacity. *Journal of Applied Physiology*, 95, 2152–2162.

Tanaka, H., & Seals, D.R. (2008). Endurance exercise performance in Masters athletes: age-associated changes and underlying physiological mechanisms. *Journal of Physiology*, 586 (Part 1), 55–63.

Volpi, E., Nazemi, R., & Fujita, S. (2004). Muscle tissue changes with aging. *Current Opinion in Clinical Nutrition and Metabolic Care*, 7, 405–410.

Weir, P.L., Kerr, T., Hodges, N.J., McKay, S.M., & Starkes, J.L. (2002). Master swimmers: How are they different from younger elite swimmers? An examination of practice and performance patterns. *Journal of Aging and Physical Activity*, 10, 41–63.

WHO Study Group (1993). *Aging and working capacity*. Geneva, Switzerland: World Health Organization.

Wilmore, J.H., & Costill, D.L. (2004). *Physiology of sport and exercise* (3rd ed.) Champaign, IL: Human Kinetics.

Wright, V.J., & Perricelli, B.C. (2008). Age-related rates of decline in performance among elite senior athletes. *American Journal of Sports Medicine*, 36, 443–450.

CHAPTER FOUR

THE EFFECTS OF AGING AND SUSTAINED EXERCISE INVOLVEMENT ON CARDIOVASCULAR FUNCTION IN OLDER PERSONS

STEVEN A. HAWKINS

Of the physiological changes associated with advancing age, those occurring in the cardiovascular (CV) system are among the most clinically and functionally relevant. Aging is associated with mild left ventricular hypertrophy, narrowing of the outflow tract from the left ventricle, reduced responsiveness to b-adrenergic stimuli, significant reduction in peak cardiac output, and impaired distensibility of arteries (Ferrari et al., 2003). In addition, aging results in diminished mitochondrial content and capacity in skeletal muscle (Conley et al., 2007). These changes dramatically limit the functional capacity of the CV system at advanced age, such that by age 75 years, over half of the maximal oxygen consumption (VO_2max) has been lost (Barnard et al., 1979). Older adults commonly demonstrate VO_2max values that are lower than required for many common activities of daily living (Durstine & Moore, 2003). VO_2max represents the functional limit of the body's ability to deliver and extract oxygen to meet the metabolic demands of vigorous physical activity, and is recognized as the international reference standard for physical fitness (Shephard et al., 1968). As well as presenting challenges to physical independence and quality of life, low cardiorespiratory fitness has been consistently associated with CV disease and all-cause mortality (Blair et al., 1996; Myers et al., 2002; Paffenbarger et al., 1978). More importantly, these studies have demonstrated low cardio-respiratory fitness to be a strong and independent predictor of mortality risk (Paffenbarger et al., 1978; Wei et al., 1999).

Beginning with the work of Robinson (1938) over 50 years ago, tremendous interest has been directed towards describing the age-related change in VO_2max. While current literature ascribes a mostly linear five to ten per cent per decade loss rate in VO_2max that begins in the third decade of life, there is growing

52

evidence that this may be incorrect. Moreover, recent work in Master Athletes has also challenged common beliefs about loss rates in this highly active segment of older adults. This chapter will review the current state of the science in relation to aging and VO$_2$max, focusing on the scope and shape of age-related loss in Master Athletes and the mechanisms that have been proposed to explain these losses.

SEDENTARY LOSS RATES

The original work by Robinson (1938), using a cross-sectional design, produced loss rates in VO$_2$max with age in men that approximated ten per cent per decade. While the study of women did not begin until sometime later, reported relative age-related loss rates in VO$_2$max in women were similar (Astrand, 1960). This value (ten per cent decline per decade) has been consistently reproduced in sedentary and moderately physically active men and women on a worldwide basis in cross-sectional studies that have followed (Table 4.1).

Additionally, longitudinal studies, which are considered more valid means of assessing age-related changes in physiological function (Dehn & Bruce, 1972), have generally reported loss rates of approximately ten per cent per decade in sedentary people, at least when subjects are between the ages of 20 and 60 years (Astrand et al., 1973; Marti & Howald, 1990; Plowman et al., 1979; Rogers et al., 1990; Trappe et al., 1996). In contrast to cross-sectional designs, older individuals tend to exhibit higher loss rates (>15 per cent per decade) in longitudinal designs (Table 4.2; Eskurza et al., 2002; Katzel et al., 2001). One notable exception is the work of Kasch et al. (Kasch et al., 1999; Kasch et al., 1995; Kasch et al., 1990; Kasch & Wallace, 1976), reporting on a group of participants involved in a moderate exercise program for up to 32 years. These authors reported loss rates that were significantly lower than the majority of studies; for example, at the ten-year follow-up, loss rates were statistically nonsignificant. At subsequent follow-ups after 22, 28 and 33 years, loss rates averaged five to seven per cent per decade. However, these loss rates were diluted by a training effect (i.e., improvements in fitness due to increased training), change in body weight, and sampling bias, so it is difficult to attribute the findings solely to physical activity.

A central concept in these findings, and one that is widely accepted, is the constant decline in VO$_2$max across the age span; that is, the linearity of loss. However, this concept has been challenged since 1987 when Buskirk and Hodgson (1987) proposed a curvilinear loss rate in VO$_2$max primarily related

the effects of aging and sustained exercise on cardiovascular function

Table 4.1 Cross-sectional rates of change in VO_2max and HR max in sedentary, active and athletic samples. N = number of subjects; YA = young athletes; MA = Master Athletes; NA = not available; W = women; M = men (reprinted with permission from Hawkins & Wiswell, 2003)

Reference	N	Gender	Age (yrs)	Groups	VO_2max decline/ decade	HRmax decline/ decade	Disease excluded
Pollock et al. (1974)	25	M	40–75 grouped by decade	MA	10%	2.5%	Yes
Drinkwater et al. (1975)	122	F	20–68 grouped by decade	Mixed activity	10%	NA	Yes
Barnard et al. (1979)	13	M	40–78 regression	MA	11.5%	5%	Yes
Heath et al. (1981)	16	M	59±6	MA	4%	4%	Yes
	16	M	22±2	YA			
Hossack & Bruce (1982)	98	M	20–75 regression	Sedentary	10% M	3% W	Yes
	104	W		9% W	5% M		
Fuchi et al. (1989)	55	M	30–80 grouped by decade	MA	7%	3%	Yes
Inbar et al. (1994)	1424	M	20–70 grouped by decade	Sedentary	7%	3%	Yes
Toth et al. (1994)	378	M	17–80	Mixed activity	8% M	NA	Yes
	224	W	18–81	10% F			
Jackson et al. (1995)	1499	M	25–70 grouped by decade	Mixed activity	10%	4%	Yes
Jackson et al. (1996)	409	W	20–64 grouped by decade	Mixed activity	10%	4%	Yes
Cunningham et al. (1997)	124	M	55–86 regression	Mixed activity	11% M	NA	Yes
	97	W			11% F		
Tanaka et al. (1997)	84	W	20–75 grouped by decade	MA	9.7% MA	3% MA	Yes
	72	W		Sedentary	9.1% Sed	3% Sed	
Rosen et al. (1998)	276	M	45–80 regression	Mixed activity	9% MA	NA	Yes
					9% Sed		
Wiebe et al. (1999)	23	W	20–63 grouped by age	MA	9%	5%	No
Schiller et al. (2001)	146	W	20–75 grouped by decade	Mixed activity	9%	4.5%	Yes
Wiswell et al. (2001)	146	M	40–86 regression	MA	12% M	NA	Yes
	82	W			8% W		
Pimental et al. (2003)	153	M	20–75 regression	MA	11% Sed	3.5% MA	Yes
				Sedentary	11% MA	4% Sed	

Table 4.2 Longitudinal rates of change in VO$_2$max and HR max in sedentary, active and athletic samples. N = number of subjects; YA = young athletes; MA = Master Athletes; Sed = sedentary; AO = active older; NA = not available; W = women; M = men (reprinted with permission from Hawkins & Wiswell, 2003)

Reference	N T1	N T2	Gender	Age (yrs) test one	Exercise Status T1	Follow-up (yrs)	Exercise Status T2	VO$_2$max decline/ decade	HRmax decline/ decade	Disease excluded
Astrand et al. (1960)	86	66	M & W	20–33	Active	21 years	Active	9% M 11% W	3% M 4% W	Yes
Robinson et al. (1975)	89	37	M	18–22	Mixed activity	31 years	Mixed activity	8%	3%	No
Kasch & Wallace (1976)	NA	16	M	32–56	Active	10 years	Active	2%	4%	Yes
Robinson et al. (1976)	16	13	M	20–54	YA	32 years	Mixed activity	13%	1%	No
Plowman et al. (1979)	122	36	W	18–68	Mixed activity	6.1 years	Mixed activity	10%		Yes
Pollock et al. (1987)	25	22	M	40–75	MA	10 years	MA Active	1% 12.5%	4% 4%	Yes
Rogers et al. (1990)	29	29	M	47–84	MA Sed	8 years	MA Sed	5.5% MA 12% Sed	6% MA 0% Sed	Yes
Marti & Howald (1990)	53	50	M	26.7±4.3 19.7±3.3	YA Sed	15 years	Athletic Sed	8% Athletic 11% Sed	3% Athletic 5% Sed	Yes
Kasch et al. (1990)	NA	30	M	48±7	Active	23 years 18 years	Active Sed	7% Active 21% Sed	NA	No
Kasch et al. (1995)	NA	24	M	33–57	Active	28 years	Active Sed	5% Active 19% Sed	5% Active 8% Sed	No
Trappe et al. (1996)	>100	53	M	18–55	YA MA	22 years	Athletic Active Sed Active older	6% Athletic 10% Active 15% Sed 15% AO	3% Athletic 3% Active 2% Sed 5% AO	Yes
Hagerman et al. (1996)	9	9	M	23.8±1.5	YA	20 years	Mixed activity	15%	7%	Yes
Pollock et al. (1997)	25	21	M	40–75	MA	20 years	MA Active Sed	15% MA 14% Active 34% Sed	4% MA 4% Active 4% Sed	No
Kasch et al. (1999)	NA	11	M	33–56	Active	33 years	Active	7%	5%	No
Katzel et al. (2000)	160	89	M	50–79	MA Sed	9 years	MA Sed	29% MA 15% Sed	4%	No
Hawkins et al. (2001)	228	135	M & W	40–86	MA	8.5 years	MA	24.5% M 11% W	4% M 5% W	Yes
Eskurza et al. (2002)	24	24	W	40–78	MA Sed	7 years	MA Sed	18% MA 15% Sed	5% MA 5% Sed	Yes

to reductions in physical activity and exercise. The authors proposed that VO_2max in sedentary individuals declines rapidly in the 20s and 30s due primarily to decreasing physical activity and increasing body mass. After this period of rapid loss, the rate slows as individuals age. Recent work from Fleg et al. (2005) has also challenged the notion of linearity in VO_2max loss rates, and further clarified the pattern of loss across the age span. The authors tracked 375 women and 435 men aged between 21 and 87 for an average of 7.9 years. All of the subjects were screened for clinical heart disease at baseline and follow-up. Fleg et al.'s (2005) data suggested loss rates of three per cent to six per cent per decade in the 20s and 30s, that increased progressively each decade to >20 per cent after age 70.

MASTER ATHLETE LOSS RATES

Efforts to define age-related loss rates are complicated by the multiple factors that influence functional decline in physiological systems, including aging, disuse, and disease (Bortz & Bortz, 1996), as well as study design issues. It has been proposed that disuse is the most significant confounding factor (Lazarus & Harridge, 2007), and that the Master Athlete represents the best model for describing inherent aging (Bortz & Bortz, 1996; Lazarus & Harridge, 2007). Early studies of athletic populations generally produced loss rates in VO_2max that were lower than ten per cent per decade. Shephard (1966), in a compilation of literature from throughout the world to calculate loss rates with age, suggested that athletic males demonstrated a loss rate of six per cent per decade. Pollock et al.'s (1974) cross-sectional study of championship athletes also described reduced loss rates, as the ten per cent decline in the entire sample was inflated by a 22 per cent per decade loss in athletes over 70 years of age. Those athletes under age 70 years demonstrated a six per cent per decade loss rate in VO_2max. Three other early studies suggested reduced loss rates in VO_2max in athletes ranging from five per cent to seven per cent (Fuchi et al., 1989; Heath et al., 1981; Meredith et al., 1987). These data have been widely interpreted to suggest that vigorous exercise reduces age-related loss of VO_2max by 50 per cent.

Additional support for reduced loss rates in Masters Athletes has been provided by longitudinal research. Several studies reported loss rates ranging from one to six per cent per decade (Dehn & Bruce, 1972; Marti & Howald, 1990; Pollock et al., 1987; Rogers et al., 1990; Trappe et al., 1996). One of these utilized a relatively short follow-up (2.3 years) in middle-aged men and described a three-fold lower rate of loss in active versus sedentary subjects

steven a. hawkins

(Dehn & Bruce, 1972). The study by Trappe et al. (1996) demonstrated reduced loss rates over a much longer time frame (22 years), but only in the younger athletes (45.3+2.3 years at follow-up). The older athletes in the study (68.4+2.7 years at follow-up) experienced loss rates of 15 per cent per decade. Similarly, the result of Marti and Howald (1990) reflected only the younger athletes (41.7+4.3 years at follow-up) who had maintained high-intensity training; these were only a very few (n = 5) of the subjects.

Reduced loss rates in older athletes have been reported by Rogers et al. (1990), and in the ten-year follow-up by Pollock et al. (1987). In the former, a comparison with sedentary men demonstrated a 50 per cent reduced loss rate in older athletes over ten years. This result, while compelling, is based on small sample size. Pollock demonstrated no loss in VO_2max in athletes who maintained training intensity and volume over the ten-year follow-up, whereas athletes who decreased training volume by approximately 15 per cent and training pace by nearly one minute/mile experienced a 13 per cent loss over the decade. This result is intriguing as the lack of change in VO_2max over time is not explainable by either an increase in training volume/intensity or change in body mass. Results were quite different when Pollock et al. reported a second follow-up after 20 years in these athletes. In those athletes who had continued to train diligently (albeit at lower levels), the total loss in VO_2max over the 20 year period was 23 per cent. Athletes who had become sedentary during the second decade of follow-up had experienced a precipitous drop of 34 per cent over that ten-year period, clearly not linear. Interestingly, work by Hagerman et al. (1996) also reported non-linear loss rates in older athletes associated with dramatic changes in training. The authors examined elite rowers for 20 years following their athletic careers, during which they were involved in mixed activity levels that were clearly reduced in volume and intensity. During the initial ten years, the athletes experienced a 20 per cent decline in VO_2max, which normalized to ten per cent in the second decade. These results support the hypothesis described earlier of non-linear loss rates in VO_2max (Buskirk & Hodgson, 1987; Fleg et al., 2005).

It is clearly an attractive hypothesis that intense exercise can reduce the loss rate in physiological function with advancing age. However, results have not consistently supported this notion, including several recent cross-sectional (Pimental et al., 2003; Tanaka et al., 1997; Wiebe et al., 1999; Wiswell et al., 2001) and longitudinal (Eskurza et al., 2002; Hawkins et al., 2001; Katzel et al., 2001) investigations. A harbinger of these results was produced by Barnard et al. (1979), where relative loss rates in VO_2max were similar for sedentary and physically active individuals. The three longitudinal reports cited above, published within a two-year time span, greatly aided our understanding of the effect of intense exercise on age-related loss of VO_2max. These studies reported loss rates ranging from 18–29 per cent per decade in the entire sample, rates two to three times greater than reported for sedentary individuals. However, in subsets of men who did not reduce training, loss of VO_2max was significantly lower, averaging 5.8 per cent over seven years in the work of Katzel and non-significant over 8.5 years in the work of Hawkins. Importantly, in the latter, these were also the men who best maintained lean body mass. These findings are similar to previously described work (see Pollock et al., 1987) demonstrating that loss in VO_2max can be reduced if training intensity is maintained. However, as seen from the 20-year follow-up of Pollock and these more recent longitudinal investigations, maintaining training volume/intensity for long periods of time appears extremely difficult, perhaps due directly to aging as well. For women, the results were not as clear. Eskurza identified a subgroup of women who did not change training patterns and yet experienced loss rates equivalent to sedentary controls. In contrast, Hawkins identified women who did not change training volume and maintained estrogen status as those best able to maintain VO_2max. Perhaps this latter point explains the difference between the two results.

Explanations for the conflicting results include small sample sizes, limited age ranges and the lack of a sedentary control group (Tanaka et al., 1997). Many of the early studies had small sample sizes, which predisposes the results to selection bias. The dissimilarities noted could also be related to training duration. In one early study demonstrating >10 per cent loss rates per decade in VO_2max in athletic adults, the authors reported that half the individuals had trained for more than 25 years consecutively (Barnard et al., 1979). Similarly, Wiswell et al. (2001) reported greater than 15 years as the average training duration with a large range and standard deviation. Therefore, a significant number of those individuals had been training for over 20 years. It has been suggested that maintaining training volume and intensity is extremely difficult for periods over ten years (Pollock et al., 1997). In addition, Hawkins et al. (2001) demonstrated the greatest reductions in training volume and intensity in the oldest (>70 years) athletes. Therefore, the studies demonstrating VO_2max loss rates of >10 per cent in athletic individuals very likely reflect periods of decreased training in addition to aging (see Weir et al., 2002).

Several concluding observations can be made concerning age-related loss in VO_2max. It appears likely that the loss rate is not linear, and while ten per cent per decade may reflect the average loss across an entire life span, it does not accurately describe the loss rate experienced in any one decade. As long

steven a. hawkins

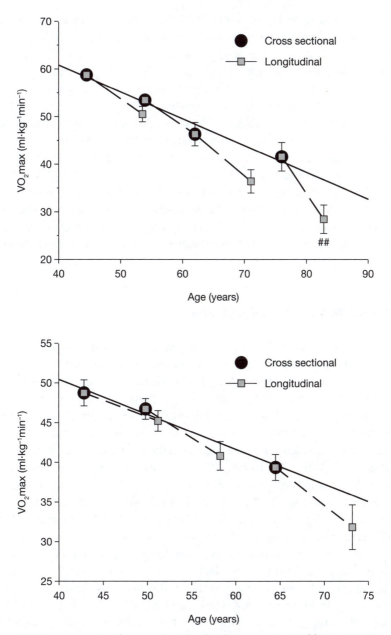

Figure 4.1 Cross-sectional versus longitudinal comparison of loss rates in maximal oxygen consumption (VO$_2$max) [ml·kg^{-1}min^{-1}] in men (upper panel) and women (lower panel) Masters Athletes (reprinted with permission from Hawkins & Wiswell, 2003)

Note: ## indicates significantly different rate of loss compared with cross-sectional

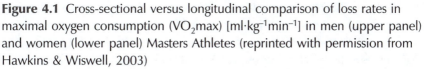

59

as physical activity and exercise levels do not change dramatically, VO_2max declines slowly from young adulthood, accelerating through middle age until a fairly dramatic loss is experienced at old age (see Figure 4.1). Compounding this age-related loss is a rapid decline at any age with reductions in physical activity and/or exercise. Moreover, it appears that high-intensity exercise training such as that undertaken by Masters Athletes can reduce loss rates in VO_2max in young and middle-aged athletes. However, this appears limited to situations in which training volume and training intensity are maintained fairly consistently. It is uncertain if older athletes can alter the age-related loss, and it appears that the ability to limit loss of VO_2max may be sustainable for only about ten years. Why this may be true isn't yet clear. However, it is important to point out that, whatever the loss rate, Masters Athletes at any age are superior to sedentary counterparts in the functional capacity of their cardiovascular system, and very likely enjoy greater cardiovascular health than those same sedentary counterparts.

MECHANISM OF LOSS OF VO_2MAX

While we don't fully understand the relative contribution of reduced exercise training and aging to the overall loss of VO_2max, it is clear that age-related loss is a combination of the two factors. Our understanding of this issue is limited by the fact that, while Masters Athletes train hard, they do train considerably fewer hours per week than young athletes (Weir et al., 2002). Early evidence suggested that reductions in VO_2max with aging were due solely to central adaptations in CV function, in particular reductions in maximal heart rate (HRmax) (Hagberg et al., 1985; Heath et al., 1981). HRmax declines at a rate uninfluenced by exercise training or gender of approximately three to five per cent per decade (Hawkins et al., 2001; Heath et al., 1981; Wiebe et al., 1999). However, the decline in HRmax is clearly linear (Fleg et al., 2005), accelerating little, if at all, across the age span. While clearly a contributor to age-related decline in VO_2max, the different patterns of loss make it unlikely that HRmax explains all or perhaps even most of the decline. There is also a decline in maximal stroke volume with age (Hagberg et al., 1985) that may (Rivera et al., 1989; Wiebe et al., 1999) or may not (Hagberg et al., 1985; Heath et al., 1981) be reduced in Masters Athletes. However, stroke volume remains adaptable in older adults as the Starling mechanism operates to increase stroke volume in response to exercise training (Rivera et al., 1989). Estimates of the contribution of central mechanisms for reduced VO_2max range from 40–100 per cent in various investigations (Hagberg et al., 1985; Heath et al., 1981; Rivera et al., 1989).

steven a. hawkins

Fleg et al. (2005) demonstrated a pattern of change in O_2 pulse (VO_2/HR) that mirrors the age-related decline in VO_2max. This measure implies a more peripheral mechanism for age-related changes involving skeletal muscle mass. This could in large part be reflective of the age-related loss of lean body mass and increase in body fat (Fleg & Lakatta, 1988; Hawkins et al., 2001; Proctor & Joyner, 1997; Toth et al., 1994). Toth et al. (1994) demonstrated that loss rates in VO_2max in sedentary men and women were reduced by 50 per cent when data were adjusted for declines in lean body mass. As well, our lab demonstrated no change in VO_2max in older athletes who maintained lean body mass over nearly ten years (Hawkins et al., 2001). It has been estimated by statistical modeling that 35 per cent of the decline in VO_2max with aging is due to age-associated declines in lean body mass (Rosen et al., 1998). However, as the pattern of age-related loss of skeletal muscle is different from the age-related loss of VO_2max, similar to reported for HRmax, it is unlikely that loss of lean body mass alone can account for all of the peripheral adaptations responsible for declining fitness with age. There may be changes in oxygen delivery or utilization occurring as well. In support of this are recent data describing a decline in mitochondrial quality with aging in skeletal muscle (Conley et al., 2000). Subjects in this study were carefully screened for disease and were physically active. Thus, the finding of reduced mitochondrial number, as well as significant declines in the capacity of the remaining mitochondria, suggests a normal process of aging that would clearly limit VO_2max.

DIRECTIONS FOR FUTURE RESEARCH

Age-related loss of VO_2max occurs in a non-linear fashion across the lifespan at an average rate of ten per cent per decade. High-intensity exercise training that is maintained across time appears able to ameliorate this loss to some degree. However, it appears that certain factors, which include aging, motivation, injury, and others, work to limit the ability to maintain training for periods much longer than one decade in most individuals. It will be important to continue to describe the age-related changes in VO_2max in athletic individuals to clarify the ability of chronic exercise to limit declines in functional capacity with advancing age. As well, it will be important to define the factors that lead to reduced training adherence with aging, and to identify the training programs that most effectively limit the age-related loss of VO_2max. For example, little research has focused on the combined effects of resistance and endurance exercise over time in relation to loss of aerobic capacity. The decline in VO_2max is due to both central and peripheral adaptations, primarily loss of lean body mass and decreases in HRmax. Changes in muscle mitochondrial function have also been implicated. Future research should focus on clarifying the role of both central and peripheral factors in the loss of VO$_2$max with aging, focusing particularly on cellular changes that might be responsible for loss of aerobic capacity and the ability of chronic exercise to influence these cellular changes.

REFERENCES

Astrand, I. (1960). Aerobic capacity in men and women with specific reference to age. *Acta Physiologica Scandinavia*, 49, 1–92.

Astrand, I., Astrand, P.-O., Hallback, I., & Kilbom, A. (1973). Reduction in maximal oxygen uptake with age. *Journal of Applied Physiology*, 35, 649–654.

Barnard, R.J., Grimditch, G.K., & Wilmore, J.H. (1979). Physiological characteristics of sprint and endurance masters runners. *Medicine and Science in Sports and Exercise*, 11, 167–71.

Blair, S.N., Kampert, J.B., Kohl, H.W., III, Barlow, C.E., Macera, C.A., Paffenbarger, R.S. Jr., & Gibbons, L.W. (1996). Influences of cardiorespiratory fitness and other precursors on cardiovascular disease and all-cause mortality in men and women. *Journal of the American Medical Association*, 276, 205–210.

Bortz, W.M., IV, & Bortz, W.M., II (1996). How fast do we age? Exercise performance over time as a biomarker. *Journal of Gerontology*, 51A, M223–225.

Buskirk, E.R., & Hodgson, J.L. (1987). Age and aerobic power: the rate of change in men and women. *Federation Proceedings*, 46, 1824–1829.

Conley, K.E., Jubrias, S.A., & Esselman, P.C. (2000). Oxidative capacity and ageing in human muscle. *Journal of Physiology*, 526, 203–210.

Conley, K.E., Jubrias, S.A., Amara, C.E., & Marcinek, D.J. (2007). Mitochondrial dysfunction: Impact on exercise performance and cellular aging. *Exercise and Sport Sciences Reviews*, 35, 43–49.

Cunningham, D.A., Paterson, D.H., Koval, J.J., & St Croix, C.M. (1997). A model of oxygen transport capacity changes for independently living older men and women. *Canadian Journal of Applied Psychology*, 22, 439–453.

Dehn, M.M., & Bruce, R.A. (1972). Longitudinal variations in maximal oxygen intake with age and activity. *Journal of Applied Physiology*, 33, 805–807.

Drinkwater, B.L., Horvath, S.M., & Wells, C.L. (1975). Aerobic power in females, ages 10 to 68. *Journal of Gerontology*, 30A, 385–394.

Durstine, J.L., & Moore, G.E. (2003). *ACSM's exercise management for persons with chronic diseases and disabilities*. Champaign, IL: Human Kinetics.

Eskurza, I., Donato, A.J., Moreau, K.L., Seals, D.R., & Tanaka, H. (2002). Changes in maximal aerobic capacity with age in endurance-trained women: 7-year follow-up. *Journal of Applied Physiology*, 92, 2303–2308.

Ferrari, A.U., Radaelli, A., & Centola, M. (2003). Invited review: Aging and the cardiovascular system. *Journal of Applied Physiology*, 95, 2591–2597.

Fleg, J.L., & Lakatta, E.G. (1988). Role of muscle mass in the age-associated reduction in VO$_2$max. *Journal of Applied Physiology*, 65, 1147–1151.

62

steven a. hawkins

Fleg, J.L., Morrell, C.H., Bos, A.G., Brant, L.J., Talbot, L.A., Wright, J.G., & Lakatta, E.G. (2005). Accelerated longitudinal decline of aerobic capacity in healthy older adults. *Circulation*, 112, 674–682.

Fuchi, T., Iwaoka, K., Higuchi, M., & Kobayashi, S. (1989). Cardiovascular changes associated with decreased aerobic capacity and aging in long distance runners. *European Journal of Applied Physiology*, 58, 884–889.

Hagberg, J.M., Allen, W.K., Seals, D.R., Hurley, B.F., Ehsani, A.A., & Holloszy, J.O. (1985). A hemodynamic comparison of young and older endurance athletes during exercise. *Journal of Applied Physiology*, 58, 2041–2046.

Hagerman, F.C., Fielding, R.A., Fiatarone, M.A., Gault, J.A., Kirkendall, D.T., Ragg, K.E., et al. (1996). A 20-year longitudinal study of Olympic oarsmen. *Medicine and Science in Sports and Exercise*, 28, 1150–1156.

Hawkins, S.A., Marcell, T.J., Jaque, S.V., & Wiswell, R.A. (2001). A longitudinal assessment of change in VO_2max and maximal heart rate in master athletes. *Medicine and Science in Sports and Exercise*, 33, 1744–1750.

Hawkins, S.A., & Wiswell, R.A. (2003). Rate and mechanism of maximal oxygen consumption decline with aging. *Sports Medicine*, 33, 877–888.

Heath, G.W., Hagberg, J.M., Ehsani, A.A., & Holloszy, J.O. (1981). A physiological comparison of young and older endurance athletes. *Journal of Applied Physiology*, 51, 634–640.

Hossack, K.F., & Bruce, R.A. (1982). Maximal cardiac function in sedentary normal men and women: comparison of age-related changes. *Journal of Applied Physiology*, 53, 799–204.

Inbar, O., Orzen, A. Scheinowitz, M., Rotstein, A., Dlin, R., & Casaburi, R. (1994). Normal cardiopulmonary responses during incremental exercise in 20- to 70-yr-old men. *Medicine and Science in Sports and Exercise*, 26, 538–546.

Jackson, A.S., Beard, E.F., Wier, L.T., Ross, R.M., Stuteville, J.E., & Blair, S.N. (1995). Changes in aerobic power of men, ages 25–70 yr. *Medicine and Science in Sports and Exercise*, 27, 113–120.

Jackson, A.S., Wier, L.T., Ayers, B.W., Beard, E.F., Stuteville, J.E., & Blair, S.N. (1996). Changes in aerobic power of women, ages 20–64 yr. *Medicine and Science in Sports and Exercise*, 28, 884–891.

Kasch, F., & Wallace, J.P. (1976). Physiological variables during 10 years of endurance exercise. *Medicine and Science in Sports and Exercise*, 8, 5–8.

Kasch, F.W., Boyer, J.L., Van Camp, S.P., Verity, L.S., & Wallace, J.P. (1990). The effect of physical activity and inactivity on aerobic power in older men (a longitudinal study). *The Physician and Sportsmedicine*, 18, 73–83.

Kasch, F.W., Boyer, J.L., Van Camp, S., Nettl, F., Verity, L.S., & Wallace, J.P. (1995). Cardiovascular changes with age and exercise: a 28-year longitudinal study. *Scandinavian Journal of Medicine and Science in Sports*, 5, 147–51.

Kasch, F.W., Boyer, J.L., Schmidt, P.K., Wells, R.H., Wallace, J.P., Verity, L.S., Guy, H., & Schneider, D. (1999). Ageing of the cardiovascular system during 33 years of aerobic exercise. *Age and Ageing*, 28, 531–536.

Katzel, L.I., Sorkin, J.D., & Fleg, J.L. (2001). A comparison of longitudinal changes in aerobic fitness in older endurance athletes and sedentary men. *Journal of the American Geriatric Society*, 49, 1657–1664.

Lazarus, N.R., & Harridge, S.D.R. (2007). Inherent aging in humans: The case for studying master athletes. *Scandinavian Journal of Medicine & Science in Sports*, 17, 461–463.

Marti, B., & Howald, H. (1990). Long-term effects of physical training on aerobic capacity: Controlled study of former elite athletes. *Journal of Applied Physiology*, 69, 1451–1459.

Meredith, C.N., Zackin, M.J., Frontera, W.R., & Evans, W. (1987). Body composition and aerobic capacity in young and middle-aged endurance-trained men. *Medicine and Science in Sports and Exercise*, 19, 557–563.

Myers, J., Prakash, M., Froelicher, V., Do, D., Partington, S., & Atwood, J.E. (2002). Exercise capacity and mortality among men referred for exercise testing. *New England Journal of Medicine*, 346, 793–801.

Paffenbarger, R.S., II, Wing, A.L., & Hyde, R.T. (1978). Physical activity as an index of heart attack risk in college alumni. *American Journal of Epidemiology*, 108, 161–175.

Pimental, A.E., Gentile, C.L., Tanaka, H., Seals, D.R., & Gates, P.E. (2003). Greater rate of decline in maximal aerobic capacity with age in endurance-trained than in sedentary men. *Journal of Applied Physiology*, 94, 2406–2413.

Plowman, S.A., Drinkwater, B.L., & Horvath, S.M. (1979). Age and aerobic power in women: a longitudinal study. *Journal of Gerontology*, 34, 512–520.

Pollock, M.L., Miller, H.S., & Wilmore, J. (1974). Physiological characteristics of champion American track athletes 40 to 75 years of age. *Journal of Gerontology*, 29, 645–649.

Pollock, M.L., Foster, C., Knapp, D., Rod, J.L., & Schmidt, D.H. (1987). Effect of age and training on aerobic capacity and body composition of master athletes. *Journal of Applied Physiology*, 62, 725–731.

Pollock, M.L., Mengelkoch, L.J., Graves, J.E., Lowenthal, D.T., Limacher, M.C., Foster, C., & Wilmore, J.H. (1997). Twenty-year follow-up of aerobic power and body composition of older track athletes. *Journal of Applied Physiology*, 82, 1508–1516.

Proctor, D.N., & Joyner, M.J. (1997). Skeletal muscle mass and the reduction of VO_2max in trained older subjects. *Journal of Applied Physiology*, 82, 1411–1415.

Rivera, A.M., Pels, A.E., III, Sady, S.P., Sady, M.A., Cullinane, E.M., & Thompson, P.D. (1989). Physiological factors associated with the lower maximal oxygen consumption of master runners. *Journal of Applied Physiology*, 66, 949–954.

Robinson, S. (1938). Experimental studies of physical fitness in relation to age. *Arbeitsphysiologie*, 10, 251–323.

Robinson, S., Dill, D.B., Tzankoff, S.P., Wagner, J.A., & Robinson, R.D. (1975). Longitudinal studies of aging in 37 men. *Journal of Applied Physiology*, 38, 263–267.

Robinson, S., Dill, D.B., Robinson, R.D., Tzankoff, S.P., & Wagner, J.A. (1976). Physiological aging of champion runners. *Journal of Applied Physiology*, 41, 46–51.

Rogers, M.A., Hagberg, J.M., Martin, W.H., III, Ehsani, A.A., & Holloszy, J.O. (1990). Decline in VO_2max with aging in master athletes and sedentary men. *Journal of Applied Physiology*, 68, 2195–2199.

Rosen, M.J., Sorkin, J.D., Goldberg, A.P., Hagberg, J.M., & Katzel, L.I. (1998). Predictors of age associated decline in maximal aerobic capacity: a comparison of four statistical models. *Journal of Applied Physiology*, 84, 2163–2170.

64

Schiller, B.C., Casas, Y.G., DeSouza, C.A., & Seals, D.R. (2001). Maximal aerobic capacity across age in healthy Hispanic and Caucasian women. *Journal of Applied Physiology*, 91, 1048–1054.

Shephard, R.J. (1966). World standards of cardiorespiratory performance. *Archives of Environmental Health*, 13, 664–670.

Shephard, R.J., Allen, C., Benade, A.J., Davies, C.T., Di Prampero, P.E., Hedman, R., Merriman, J.E., Myhre, K., & Simmons, R. (1968). The maximal oxygen intake: an international reference standard of cardiorespiratory fitness. *Bulletin of the World Health Organization*, 38, 757–764.

Tanaka, H., DeSouza, C.A., Jones, P.P., Stevenson, E.T., Davy, K.P., & Seals, D.R. (1997). Greater rate of decline in maximal aerobic capacity with age in physically active vs sedentary healthy women. *Journal of Applied Physiology*, 83, 1947–1953.

Toth, M.J., Gardner, A.W., Ades, P.A., & Poehlman, E.T. (1994). Contribution of body composition and physical activity to age-related decline in peak VO_2 in men and women. *Journal of Applied Physiology*, 77, 647–652.

Trappe, S.W., Costill, D.L., Vukovich, M.D., Jones, J., & Melham, T. (1996). Aging among elite distance runners: a 22-yr longitudinal study. *Journal of Applied Physiology*, 80, 285–290.

Wei, M., Kampert, J.B., Barlow, C.E., Nichaman, M.Z., Gibbons, L.W., Paffenbarger, R.S. Jr., & Blair, S.N. (1999). Relationship between low cardiorespiratory fitness and mortality in normal-weight, overweight, and obese men. *Journal of the American Medical Association*, 282, 1547–1553.

Weir, P.L., Kerr, T., Hodges, N.J., McKay, S.M., & Starkes, J.L. (2002). Master swimmers: How are they different from younger elite swimmers? An examination of practice and performance patterns. *Journal of Aging & Physical Activity*, 10, 41–64.

Wiebe, C.G., Gledhill, N., Jamnik, V.K, & Ferguson, S. (1999). Exercise cardiac function in young through elderly endurance trained women. *Medicine and Science in Sports and Exercise,* 31, 684–691.

Wiswell, R.A., Hawkins, S.A., Jaque, S.V., Hyslop, D., Constantino, N., Tarpenning, K., Marcell, T., & Schroeder, E.T. (2001). Relationship between physiological loss, performance decrement, and age in master athletes. *Journal of Gerontology*, 56A, M1–9.

CHAPTER FIVE

MAINTENANCE OF SKILLED PERFORMANCE WITH AGE

Lessons from the Masters

JOSEPH BAKER AND JÖRG SCHORER

> Iron rusts from disuse, stagnant water loses its purity and in cold weather becomes frozen; even so does inaction sap the vigors of the mind.
>
> Leonardo da Vinci

Worldwide life expectancy is climbing approximately three months per year (Oeppen & Vaupel, 2002). But despite the positive consequences associated with this trend, a significant proportion of older persons will spend this extended lifespan in ill health. Considerable research, much of it summarized in other chapters of this book, indicates a regular and predictable decline in physical and cognitive performance with advancing age. Examinations of variables ranging from reaction time (Etnier et al., 2003; Fozard et al., 1994) to components of memory (Henry et al., 2004), muscular strength (Kallman et al., 1990) and flexibility (Einkauf et al., 1987) imply a downward spiral of functional ability with age.

On a more positive note, although consistent evidence indicates physical and cognitive capabilities decline *with* age, there is contradictory evidence as to whether this is actually *due to* age. Significant minorities of older persons report no functional limitations, and as a result, many gerontological researchers have focused on exploring factors associated with optimal health in later life. Maharam et al. suggested that many of the factors thought to explain physical and cognitive declines associated with aging were in fact the result of a 'long-standing sedentary lifestyle or disuse' (Maharam et al., 1999, p. 274), a conclusion reinforced by tracking studies indicating that physical activity and exercise involvement decline with age (Malina, 2001).

A primary concern for aging persons is maintenance of learned physical and cognitive skills. Encouragingly, researchers examining the development and

joseph baker and jörg schorer

maintenance of psychomotor performance have shown that this performance can be remarkably stable with age. This chapter summarizes research on the maintenance of performance in skilled tasks and provides an overview of theoretical explanations for this maintenance.

WHAT IS SKILLED PERFORMANCE?

Skilled performance refers to tasks that have been learned through experience or acquired through training — an important distinction from innate capacities such as simple reaction time or proportion of specific muscle-fiber types. It can take many forms, from brushing one's hair to reading a newspaper. This chapter discusses general areas of performance when appropriate but largely focuses on skilled performance in sport. Athletic contests are an ideal vehicle for measuring age-related decline since the comprehensive record keeping in many sports affords readily accessible measures of performance over time. Further, an examination of skilled sport performance can inform our understanding of performance in Masters-level athletes. Several researchers (e.g., Bortz & Bortz, 1996; Stones & Kozma, 1980, 1984; see Stones, Chapter 2) have used this type of data to examine the decline in performance with advancing age. For example, Bortz and Bortz (1996) examined data from a range of athletic events (among other things) and suggested that a decline of 0.5 per cent per year from peak performance represents a basic biomarker of aging. However, their analysis considered sports where performance is largely limited by factors affecting aerobic efficiency, and there is no evidence that a similar rate of decline would apply to skilled performance. It is more likely that tasks where cognitive and technical skills play a comparatively larger role are better maintained. Below, we summarize research from a range of cognitive and motor domains that provide a profile of aging incongruent with the widely accepted belief of aging as a period of *inevitable* decline. We make use of the Newell & Simon (1972) model of information processing that suggests human behavior can be broken down into three interacting stages: (a) *perceptual processes*, (b) *response selection*, and (c) *response execution,* to present our case. While this model is not without limitations (see Newell et al., 2001), it provides a useful framework for the discussion that follows.

Perceptual Processes

Specific cognitive and perceptual adaptations are of critical importance for explaining highly skilled performance. For instance, perception of meaningful information has been presented as an important variable that differentiates highly skilled from lesser skilled performers (cf. Williams & Ward, 2003). Researchers have consistently reported experts do not differ from non-experts in visual abilities, but rather, the differences lie in the experts' ability to interpret visual information more efficiently and effectively (Abernethy et al., 1994). Studies examining the utilization of visual information from expert cricket (Abernethy & Russell, 1984), badminton (Abernethy, 1991; Abernethy & Russell, 1987), soccer (Williams & Burwitz, 1993), and squash players (Abernethy, 1990; Abernethy, 1991) reveal that experts use body position cues obtained from their opponents' pre-contact movements (i.e., before they bowl, or before they strike the birdie or ball) to provide information about how best to respond in a situation. Due to the limited utility of post-opponent-contact information (i.e., after the opponent bowls or strikes), experts from these sports have learned to rely on visual cues from the opponent's pre-contact arm and wrist position to predict where their opponent will place the ball (cf. Land & McLeod, 2000). They then make the necessary adjustments to attend to their prediction.

Utilization of highly specific perceptual skills such as those used in many sports is possible only after many years of intensive practice (Ericsson et al., 1993). Although several studies have considered how this type of skilled perception is acquired (e.g., Baker et al., 2003; Ward et al., 2007), we have little understanding of how stable it is over time. Previous examinations of perceptual decline have shown age-related changes in the ability to perceive motion (Gilmore et al., 1992), speed (Norman et al., 2003), and depth (Norman et al., 2000) as well as the ability to distinguish 2-D and 3-D shapes (Andersen & Atchley, 1995; Norman et al., 2000). However, these studies have relied exclusively on perceiving information in novel tasks using non-expert samples. Research robustly indicates that skilled perception is domain-specific, and as a result, general or novel tasks may not be appropriate for investigating the influence of age on skilled perception.

A recent study by our research team suggests that certain types of skilled perception are quite resistant to decline. Schorer and Baker (2009) considered five groups of handball goalkeepers on a range of tasks designed to measure cognitive and perceptual skills related to this position. The groups ranged in age from ~14 years (active junior players) to ~47 years (inactive-retired players). Amazingly, retired handball goalkeepers performed similarly to current expert players and superior to lower level players on the skilled perception task despite being retired for an average of ten years.[1] Although these results are preliminary and require replication, they suggest that skilled perception is not vulnerable to the same rate of age-related decline as more general measures of perception

68

outlined above. The explanation for this result may lie in the nature of the capacities being compared. Measures of general perceptual abilities have typically examined participants' aptitude for perceiving novel stimuli, while skilled perception deals with individuals' skill at perceiving highly specific, domain-relevant information. Moreover, the nature of skilled perception for athletes in time-constrained, decision-making sports may be more akin to a problem-solving task. Experts from these sports have learned to rely on advanced visual information provided by their opponent, and once they have learned the necessary sources of this information, it may be highly resistant to memory decay. Similar to the old adage 'you never forget how to ride a bike', it may be that experts in some sports never forget where to look for the most information-rich visual cues. In support of this conclusion, evidence from this study suggested retired players and current players were very similar in the way they looked at a visual display (as measured by location and duration of eye fixations). These preliminary results suggest something quite interesting about skilled perception and warrant additional research using other tasks, which may relate more generally to aging populations (e.g., skilled perception while driving).

Response Selection

In addition to the expert's use of skilled perception, there is evidence that the way experts store, search, and retrieve information is different from lesser skilled individuals, and that this difference also comes after years of extensive practice and training. McPherson et al. have noted that experts in baseball (McPherson, 1993), volleyball (McPherson, 1993), and tennis (McPherson & French, 1991) have a deeper and more structured knowledge of the specific sport to access when making decisions. When placed in a decision-making situation during competitions, experts examine more possible options and have more detailed heuristics to determine their choice of action.

In addition to this deeper and more efficiently structured knowledge base, experts are better able to detect meaningful patterns of information. Research with chess experts by de Groot (1965) and Simon and Chase (Chase & Simon, 1973; Simon & Chase, 1973) noted that experts were superior to non-experts in their ability to recreate structured patterns. However, this 'advantage' was confined to patterns that had relevance to their area of expertise and disappeared when pieces were not organized in a structured offensive or defensive chess pattern (i.e., a random placement of pieces on the chessboard). Based on these results, Simon and Chase (1973) speculated that experts did not have superior memory capacity, but rather they had the ability to store more elaborate patterns or 'chunks' of domain-specific information than non-experts. The

69

implication of their research was that differences in skill level could be attributed to both the organization of memory space and the efficiency of the search of that space. Since this initial study, similar findings have been noted in bridge playing (e.g., Charness, 1979), medical diagnosis (e.g., Patel et al., 1990), computer programming (e.g., Barfield, 1986), and map reading (e.g., Howard & Kerst, 1981). Similarly, expertise in many sports seems to be underpinned by this skill (Abernethy et al., 2005; Allard et al., 1980; Borgeaud & Abernethy, 1987; Starkes, 1987).

Few studies have considered the stability of this superior information structure across the lifespan; however, the limited evidence available suggests it is more resistant to age-related mechanisms of decline. Charness (1981) examined age-related qualitative change in chess players by having them 'think aloud' while considering different chess problems. Results indicated that the quality of chess moves by older adults did not diminish with age, in spite of the fact that older experts engaged in a less extensive search process.

Response Execution

Knowledge regarding the maintenance of perceptual and response selection processes should still be viewed as incomplete. Much greater attention, however, has been given to understanding the extent to which high levels of motor performance can be maintained with age. Our research group has done several studies of skill maintenance in expert golfers (Baker et al., 2005, 2007). Golfers tend to peak later than athletes in most sports (Schulz & Curnow, 1988), suggesting performance in this endeavor may be less constrained by biological systems and more reliant on acquired skills. Moreover, elite golfers can spend a considerably extended time performing at the highest levels of play, as exemplified by Greg Norman's third-place finish in the 2008 British Open at the age of 53. In one study, we (Baker et al., 2005) examined the performance of elite professional golfers from the age of 35 through to age 60 as measured by scoring average — the average number of strokes taken to complete an 18-hole round of golf. The golfers showed a decline of just 0.07 per cent per year from age 35–50. This decline accelerated to 0.25 per cent per year from ages 51–60, although even this accelerated rate was just half of that predicted for more physiologically-constrained activities (Bortz & Bortz, 1996).

One limitation of this first study was that only one performance outcome was examined, and while scoring average is arguably the best measure of golf performance, it is a composite of other performance measures. While all golf

70

skills require a high degree of biomechanical efficiency, other factors are also important. Putting, for example, involves very fine motor proficiency, whereas driving distance requires a greater reliance on muscular power, which tends to peak earlier and decline more rapidly (Schultz & Curnow, 1988). Examination of the individual golf skills allows a more comprehensive investigation of rates of decline across a range of motor skills. To this end, we (Baker et al., 2007) examined age-related changes in performance across the following skills: (a) scoring average, (b) driving distance (average yards per drive), (c) driving accuracy (per cent of time drive lands in fairway), (d) greens hit in regulation (per cent of time player is able to land on the green in regulation, which is determined by the golfer's ability to hit greens on a par five hole in three strokes or less, greens on a par four hole in two strokes or less, and greens on a par three hole in a single stroke), and (e) putts per round (number of putts per 18-hole round of golf). Our results showed that the rate and profile of decline varied across the skills (see Table 5.1). Greens in regulation and driving distance had the greatest rates of decline compared with scoring average and putts per round. Driving accuracy, interestingly, showed a small improvement in performance over the same period. In sum, these studies of elite golfers reinforce the conclusion that performance in this sport is remarkably more stable with age than performance in biologically constrained activities such as running and swimming.

Studies in other motor domains have also found performance to be only marginally affected by age. Salthouse (1984) examined 74 typists (ages 19–72) in two separate studies on a number of different tasks designed to gauge both typing-specific skills along with more general measures of perceptual-motor and cognitive efficiency. He found that performance of typists remained virtually unchanged across the adult lifespan. Collectively, these studies support the conclusion that performance across the stages of information processing can be remarkably consistent with age.

Table 5.1 Age-related decline in golf skills from age of peak performance to 50 years

Skill	Rate of decline
Greens in regulation	−0.36
Driving distance	−0.23
Scoring average	−0.14
Putts per round	−0.11
Driving accuracy	+0.03

Note: minus (−) sign indicates loss in performance while plus (+) sign indicates increase in performance

MECHANISMS OF SKILL MAINTENANCE

Several frameworks have been proposed to explain skill maintenance, but these can be largely categorized as *compensation theories* and *experiential theories*. The basis of compensation theories is that it is possible for overall execution of a skill to remain stable with age, despite declines in some aspects of functioning, due to an increased reliance on other aspects (Figure 5.1). More simply, the theory suggests that skilled performers compensate (either consciously or subconsciously) for a decline in one skill area by developing or improving in another. Two excellent examples of this research come from the studies discussed above that considered chess players and typists. Charness (1981) found that skilled, older chess players could perform at the same level as younger skilled players despite age-related deficiencies in memory ability. He explained these results by suggesting that older players compensate for their declining memory by more efficient information processing (i.e., a more systematic search of the problem space and a better global evaluation of chess positions). If we return to Salthouse's typists, they compensated for age-related declines by scanning further ahead in the text, which allowed them to begin keystroke preparation earlier. As a result of this advanced planning, aging typists could offset a deficiency in one area by improving performance in another (similar results were noted by Bosman, 1993). Similarly, the handball goalkeepers in the Schorer and Baker (2009) study may have compensated for age-related losses in motor speed and efficiency by anticipating earlier; indeed, there were some data to support this conclusion although additional work is needed.

Recently, brain imaging studies have provided some support for the basic tenet of compensation theory — that losses in one area lead to additional reliance on other areas. Typically, these studies have relied on positron emission tomography (PET) or functional magnetic resonance imaging (fMRI) to provide brain images of performance-matched younger and older participants performing cognitive tasks. The assumption is that, since the groups are matched on

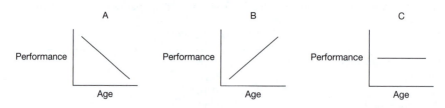

Figure 5.1 The compensation model of aging suggests that, although components of performance may decline (A), increases in a compensatory skill (B) allow for stability of performance over time (C) (from Baker, 2007)

joseph baker and jörg schorer

performance, differences in brain activation patterns between the younger and older groups would reflect the influence of age. Researchers have noted overactivation in brain regions of older participants performing perceptual, motoric, mnemonic, verbal, and spatial tasks compared to younger controls (Reuter-Lorenz & Cappell, 2008). For instance, Gutchess et al. (2005) noted that younger adults had greater activation in the medial temporal lobe than older adults when successfully performing a memory task. Conversely, successful older adults had greater activation in the prefrontal areas, suggesting that prefrontal activity may compensate for medial temporal lobe declines to support successful memory. Despite these results, it is prudent at this stage to consider this work very preliminary since it is not clear what the pattern of overactivation means. On the one hand, it could indicate a resource-related limitation such that functions previously carried out by one area must now be divided across several areas; or it could indicate a more effective organization strategy based on additional years of cognitive training. Future work, especially longitudinal studies, is necessary to determine how these patterns relate to age-related decline in general and psychological compensation in particular.

Experiential theories, such as the model of selective maintenance advocated by Ericsson et al. (Ericsson, 2000; Krampe & Ericsson, 1996), center on the notion that performance in skilled domains is maintained in very specific capacities, and these capacities can be maintained as long as practice persists. However, it seems that not just any type of practice is sufficient. Ericsson et al. (1993) proposed that engagement in 'deliberate practice' is necessary for the attainment and maintenance of expertise. Deliberate practice is a very specific type of involvement designed to improve current levels of performance requiring considerable physical and/or cognitive effort. In a review of studies on skill acquisition and learning, Ericsson (1996) concluded that, with few exceptions, level of performance was determined by the amount of time spent performing a 'well-defined task with an appropriate difficulty level for the particular individual, informative feedback, and opportunities for repetition and corrections of errors' (pp. 20–21). Hypothetically, continually modifying the level of task difficulty should allow experts to maintain (or increase) their performance by preventing learning plateaus and perpetuating adaptation to higher amounts of training stress.

To test this hypothesis, Krampe and Ericsson (1996) compared older and younger pianists on a range of performance-related measures. In addition, they compared performers at the expert and amateur levels (i.e., older expert, older amateur, younger expert, and younger amateur). They found that older performers, both amateur and expert, showed the same pattern of age-related decline on measures thought to underlie piano performance, such as reaction time; however, domain-specific measures of piano performance, such as finger tapping speed and quality

maintenance of skilled performance with age

of performance, were maintained to a greater extent in older experts. In most cases, differences in domain-specific measures of piano performance between younger and older experts were explained by differences in the amount of training and practice rather than age. Based on these results, the authors concluded that persistent regular involvement in a domain over time would allow aging performers to maintain their skills.

Despite Ericsson's position on skill maintenance being largely confined to the expert performance field, it shares some similarities with a more general model of aging proposed by Stine-Morrow (2007). Her 'Dumbledore Hypothesis'[2] is based on the notion that our choices throughout life determine our level of cognitive vitality in older age, such that, if we choose to engage in cognitively (and presumably physically) demanding activities, we will maintain these capacities to a greater degree.

Although the basic premise of experiential theories seems reasonable, some important concerns remain. For example, it is not clear whether older persons experience the same 'effect' from an event compared to their younger counterparts. It is more likely that a lifetime of experience affects an individual's response to a given stimulus (see Davids & Baker, 2007), and as a result, models of development that are appropriate for younger persons may not be applicable to the aging population. In particular, it is questionable whether older persons are able to complete the type of intense practice advocated under the deliberate-practice approach, especially in demanding sports (see Fell & Williams, Chapter 6). Moreover, simple maintenance of training does not seem to be enough to stave off age-related declines in some areas. The possibility that a threshold exists between the amount of practice necessary for performance maintenance and the absolute training limits of the aging human is an important area of future work.

The primary theoretical difference between experiential theories like Ericsson's and compensation-based theories is that experiential theories are grounded in the notion that attention to deliberate practice *prevents* age-related declines, whereas compensation theories focus on cognitive and physical mechanisms that permit performers to *compensate for* age-related declines. Despite this difference, there is likely some overlap between these two theories. For instance, domain-specific practice (as advocated by Ericsson) is likely quite similar to the domain-specific experiences of the typists in Salthouse's study, which facilitated their acquisition of compensatory mechanisms. Future work is needed to consider the extent to which these theories implicate similar qualities and, more importantly, how a more comprehensive framework could be developed through an integration of the most empirically- and conceptually-sound aspects of each.

74

CONCLUDING REMARKS

Performance in skill-based sports, which require considerable learning/practice, may be more closely aligned with precision-motor activities, such as chess and piano, than they are with sports constrained by physiological factors such as aerobic capacity. Skills that require a finer degree of motor control, necessitating a longer period of acquisition, may decline more slowly, particularly if practitioners can compensate for declining abilities through continued extensive practice (Krampe & Ericsson, 1996) or by strategically compensating for a decline in one skill area by developing or improving in another (Charness, 1981; Salthouse, 1984). Given society's increasing reliance on learned skills (particularly technological skills), the findings summarized in this chapter provide important evidence that age-related declines are not as inevitable as previously thought.

NOTES

1 Differences were found between the retired players and current players in motor proficiency.
2 The hypothesis gets its name from the advice the wizard Dumbledore gives to the title character of *Harry Potter and the chamber of secrets:* 'It is our choices . . . that show what we really are, far more than our abilities'.

REFERENCES

Abernethy, B. (1990). Anticipation in squash: Differences in advance cue utilization between expert and novice players. *Journal of Sports Sciences*, 8, 17–34.
Abernethy, B. (1991). Visual search strategies and decision-making in sport. *International Journal of Sport Psychology*, 22, 189–210.
Abernethy, B., & Russell, D.G. (1984). Advance cue utilisation by skilled cricket batsmen. *The Australian Journal of Science and Medicine in Sport*, 16, 2–10.
Abernethy, B., & Russell D.G. (1987). The relationship between expertise and visual search strategy in a racquet sport. *Human Movement Science*, 6, 283–319.
Abernethy, B., Neal, R.J., & Koning, P. (1994). Visual-perceptual and cognitive differences between expert, intermediate, and novice snooker players. *Applied Cognitive Psychology*, 8, 185–211.
Abernethy, B., Baker, J., & Côté, J. (2005). Transfer of pattern recall skills as a contributor to the development of sport expertise. *Applied Cognitive Psychology*, 19, 705–718.
Allard, F., Graham, S., & Paarsulu, M.E. (1980). Perception in sport: basketball. *Journal of Sport Psychology*, 2, 14–21.

Andersen, G.J. & Atchley, P. (1995). Age-related differences in the detection of three-dimensional surfaces from optic flow. *Psychology and Aging*, 10, 650–658.

Baker, J. (2007). Sport and physical activity in the older athlete. In P. Crocker (Ed.), *Sport psychology: a Canadian perspective* (pp. 295–314). Toronto, Ontario: Pearson.

Baker, J., Côté, J., & Abernethy, B. (2003). Sport specific training, deliberate practice and the development of expertise in team ball sports. *Journal of Applied Sport Psychology*, 15, 12–25.

Baker, J., Horton, S., Pearce, W., & Deakin, J. (2005). A longitudinal examination of performance decline in champion golfers. *High Ability Studies*, 16, 179–185.

Baker, J., Horton, S., Pearce, W., & Deakin, J. (2007). Maintenance of skilled performance with age: A descriptive examination of professional golfers. *Journal of Aging and Physical Activity*, 15, 300–317.

Barfield, W. (1986). Expert-novice differences in software: Implications for problem solving and knowledge acquisition. *Behaviour and Information Technology*, 5, 15–29.

Borgeaud, P., & Abernethy, B. (1987). Skilled perception in volleyball defence. *Journal of Sport Psychology*, 9, 400–406.

Bortz, W.M., IV, & Bortz, W.M., II (1996). How fast do we age? Exercise performance over time as a biomarker. *Journal of Gerontology: Medical Sciences*, 51A, M223–M225.

Bosman, E.A. (1993). Age-related differences in the motoric aspects of transcription typing skill. *Psychology and Aging*, 8, 87–102.

Charness, N. (1979). Components of skill in bridge. *Canadian Journal of Psychology*, 33, 1–16.

Charness, N. (1981). Search in chess: Age and skill differences. *Journal of Experimental Psychology: Human Perception and Performance*, 7, 467–476.

Chase, W.G., & Simon, H.A. (1973). The mind's eye in chess. In W.G. Chase (Ed.), *Visual information processing* (pp. 215–282). New York: Academic Press.

Davids, K., & Baker, J. (2007). Genes, environment and sport performance: Why the nature-nurture dualism is no longer relevant. *Sports Medicine*, 37, 961–980.

de Groot, A. (1965). *Thought and choice in chess*. The Hague: Mouton.

Einkauf, D.K., Gohdes, M.L., Jensen, G.M., & Jewell, M.J. (1987). Changes in spinal mobility with increasing age in women. *Physical Therapy*, 67, 370–375.

Ericsson, K.A. (1996). The acquisition of expert performance: An introduction to some of the issues. In K.A. Ericsson (Ed.), *The road to excellence: The acquisition of expert performance in arts and sciences, sports and games* (pp. 1–50). Mahwah, NJ: Erlbaum.

Ericsson, K.A. (2000). How experts attain and maintain superior performance: Implications for the enhancement of skilled performance in older individuals. *Journal of Aging and Physical Activity*, 8, 366–372.

Ericsson, K.A., Krampe, R.T., & Tesch-Römer, C. (1993). The role of deliberate practice in the acquisition of expert performance. *Psychological Review*, 100, 363–406.

Etnier, J.L., Sibley, B.A., Pomeroy, J., & Kao, J.C. (2003). Components of reaction time as a function of age, physical activity, and aerobic fitness. *Journal of Aging and Physical Activity*, 11, 319–332.

76

Fozard, J.L., Vercruyssen, M., Reynolds, S.L., Hancock, P.A., & Quilter, R.E. (1994). Age differences and changes in reaction time: The Baltimore Longitudinal Study of Aging. *Journal of Gerontology*, 49, 179–189.

Gilmore, G.C., Wenk, H.E., Naylor, L.A., & Stuve, T.A. (1992). Motion perception and aging. *Psychology and Aging*, 7, 654–660.

Gutchess, A.H., Welsh, R.C., Hedden, T., Bangert, A., Minear, M., Liu, L.L., & Park, D.C. (2005). Aging and the neural correlates of successful picture encoding: Frontal activations compensate for decreased medial-temporal activity. *Journal of Cognitive Neuroscience*, 17, 84–96.

Henry, J.D., MacLeod, M.S., Phillips, L.H., & Crawford, J.R. (2004). A meta-analytic review of prospective memory and aging. *Psychology and Aging*, 19, 27–39.

Howard, J.H., & Kerst, S.M. (1981). Memory and perception of cartographic information for familiar and unfamiliar environments. *Human Factors*, 23, 495–504.

Kallman, D.A., Plato, C.C., & Tobin, J.D. (1990). The role of muscle loss in the age-related decline of grip strength: Cross section and longitudinal perspectives. *Journal of Gerontology: Medical Sciences*, 45, M82–M88.

Krampe, R.T., & Ericsson, K.A., (1996). Maintaining excellence: Deliberate practice and elite performance in young and older pianists. *Journal of Experimental Psychology: General*, 125, 331–359.

Land, M.F., & McLeod, P. (2000). From eye movements to actions: How batsmen hit the ball. *Nature neuroscience*, 3, 1340–1345.

Maharam, L.G., Bauman, P.A., Kalman, D., Skolnik, H., & Perle, S.M. (1999). Masters athletes: Factors affecting performance. *Sports Medicine*, 28, 273–285.

Malina, R.M. (2001). Adherence to physical activity from childhood to adulthood: A perspective from tracking studies. *Quest*, 53, 346–355.

McPherson, S.L. (1993). Knowledge representation and decision-making in sport. In J.L. Starkes & F. Allard (Eds.), *Cognitive issues in motor expertise* (pp. 159–188) Amsterdam: Elsevier.

McPherson, S.L., & French, K.E. (1991). Changes in cognitive strategies and motor skills in tennis. *Journal of Sport & Exercise Psychology*, 25, 249–265.

Newell, A., & Simon, H.A. (1972). *Human problem solving*. Edgewood Cliffs, NJ: Prentice Hall.

Newell, K.M., Liu, Y-T., & Mayer-Kress, G.M. (2001). Time scales in motor learning and development. *Psychological Review*, 108, 57–82.

Norman, J.F., Dawson, T.E., & Butler, A.K. (2000). The effects of age upon the perception of depth and 3–D shape from motion and binocular disparity. *Perception*, 29, 1335–1359.

Norman, J.F., Ross, H.E., Hawkes, L.M., & Long, J.R. (2003). Aging and the perception of speed. *Perception*, 32, 85–96.

Oeppen, J., & Vaupel, J.W. (2002). Broken limits to life expectancy. *Science*, 296, 1029–1031.

Patel, V.L., Groen, G.J., & Arocha, J.F. (1990). Medical expertise as a function of task difficulty. *Memory & Cognition*, 18, 394–406.

Reuter-Lorenz, P.A., & Cappell, K.A. (2008). Neurocognitive aging and the compensation hypothesis. *Current Directions in Psychological Science*, 17, 177–182.

Salthouse, T.A. (1984). Effects of age and skill in typing. *Journal of Experimental Psychology: General*, 113, 345–371.

Schorer, J., & Baker, J. (2009). Aging and perceptual-motor expertise in handball goalkeepers. *Experimental Aging Research*, 35, 1–19.

Schulz, R., & Curnow, C. (1988). Peak performance and age among superathletes: Track and field, swimming, baseball, tennis, and golf. *Journal of Gerontology: Psychological Sciences*, 43, 113–120.

Simon, H.A., & Chase, W.G. (1973). Skill in chess. *American Scientist*, 61, 394–403.

Starkes, J.L. (1987). Skill in field hockey: the nature of the cognitive advantage. *Journal of Sport Psychology*, 9, 146–160.

Stine-Morrow, E.A.L. (2007). The Dumbledore hypothesis of cognitive aging. *Current Directions in Psychological Science*, 16, 295–299.

Stones, M.J., & Kozma, A. (1980). Adult age trends in record running performances. *Experimental Aging Research*, 6, 407–416.

Stones, M.J., & Kozma, A. (1984). Longitudinal trends in track and field performances. *Experimental Aging Research*, 10, 107–110.

Ward, P., Hodges, N.J., Starkes, J.L., & Williams, A.M. (2007). The road to excellence: Deliberate practice and the development of expertise. *High Ability Studies*, 18, 119–153.

Williams, A.M., & Burwitz, L. (1993). Advance cue utilization in soccer. In T. Reilly, J. Clarys, & A. Stibbe (Eds.), *Science and Football II* (pp. 239–244). London: E. & F.N. Spon.

Williams, A.M., & Ward, P. (2003). Perceptual expertise: development in sport. In J.L. Starkes & K.A. Ericsson (Eds.), *Expert performance in sports: Advances in research on sport expertise* (pp. 217–247). Champaign, IL: Human Kinetics.

CHAPTER SIX

AGING AND RECOVERY
Implications for the Masters Athlete

JAMES FELL AND ANDREW WILLIAMS

There is a perception among many Masters Athletes that recovery from exercise training is progressively impaired with age. A recent high-profile example of this belief is USA Olympic swimmer Dara Torres, who, despite qualifying for the individual 100m freestyle in addition to the 50m freestyle and the 4x100m freestyle and medley relays for the Beijing Olympic games, withdrew from the event as she felt that delayed recovery from multiple events at 41 years of age would impair her chances of succeeding in the 50m and the relays (Cuda, 2008).

The term 'recovery' may have many connotations for the Masters Athlete (MA), including recovery from injury, recovery from training and competition, and even recovery from post-race partying. This review focuses on the recovery from training and competition because it is the ability of the MA to successfully manipulate the training variables of *overload* and *recovery* leading to *adaptation* that enables the achievement of peak sporting performance. Similarly, the health benefits we associate with participation in regular exercise throughout the lifespan also rely on these same mechanisms insofar as the exercise stimulus must be sufficient to provide a catalyst for tissue adaptation. This adaptation ultimately provides protection from subsequent physical or environmental challenges experienced as a consequence of both living and aging.

Aging has been strongly associated with declines in physical and functional capacity despite continued engagement in regular exercise (Galloway et al., 2002). This chapter explores some of the existing evidence that investigates whether age-related physical declines may be attributable to changes in fatigue and recovery kinetics with aging. The possibility that physical and functional losses with age may be a consequence of unavoidable *detraining* as impaired recovery processes begin to prohibit the maintenance of high levels of physical activity will also be explored.

79

The principles of overload and recovery can be narrowed down to what might occur during a single training session. Most forms of physical training lead to the progression of fatigue, which effectively reduces the capacity for peak performance. However, this mechanism is essential to the training process as it provides the stimulus for recovery processes that enable the stressed systems to adapt and improve. The goal is for the repair mechanisms to enable restoration of pre-training session performance and for adaptive mechanisms to prepare the individual to cope better with any subsequent stressor of similar nature. These processes should ultimately lead to progressive improvements in functional capacity and performance.

A basic model for such fatigue-and-recovery processes is presented in Figure 6.1 and demonstrates the exercise-related loss of function followed by the period of recovery, leading to a degree of supercompensation typical of physiological variables such as muscle glycogen levels. It is the time course of the recovery phase in this figure that is of interest for this review. Recovery kinetics have been suggested to follow a curved response, rather than a straight line, with a rapid initial phase, leveling off to a slower rate of recovery as time progresses (Bompa, 1999; Koutedakis et al., 2006). The duration required for the complete recovery process and restoration of physical performance depends upon the sport, the phase of the training or competition season, the goals of the athlete, and, potentially, age. Training-induced fatigue can be minor (such

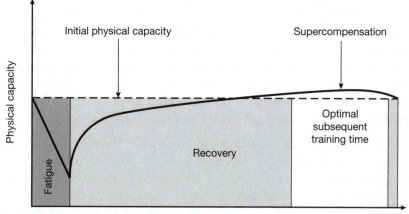

Figure 6.1 Physical capacity in response to a single bout of exercise-induced fatigue followed by a period of optimal recovery

80

as phosphocreatine [PCr] depletion) with a short recovery duration, or more long term as a consequence of factors such as exercise-induced muscle damage or neuro-endocrine disturbances. The question of interest to the aging athlete is whether the duration required for complete recovery from a given training stimulus is the same for athletes of all ages. A complicating issue is that recovery may depend on factors such as the training level and history of the athlete, the nature of the training stimulus, and the methods used to measure recovery.

There is a common belief among MAs that recovery is slowed with age. This belief is clearly evident in the abundance of literature on this topic available on the Internet. However, despite the anecdotal evidence, there has been very little research to empirically support this belief, particularly in well-trained MAs (Fell & Williams, 2008). The potential cause for a longer recovery period in the MA is probably due to either: (a) greater fatigue as a result of the same exercise bout, (b) slower rate of recovery, leading to impaired adaptation, or (c) a combination of both of these factors. Regardless of the cause, any impairment of recovery in the MA has serious implications for the training process and peak athletic performance in this population.

POTENTIAL IMPACT OF INADEQUATE RECOVERY

Coaches and athletes adopt a wide variety of approaches to manipulate the training program with the ultimate goal being to elicit maximal improvements in performance. In its simplest form, a training program should be effective when subsequent training sessions occur during the supercompensatory period from the previous exercise bout (Koutedakis et al., 2006). If training sessions are too frequent, subsequent bouts will occur before complete recovery is achieved. Alternatively, if the duration between bouts is too long, it is possible that the positive adaptations from the earlier exercise bout will have diminished. Either of these scenarios will limit the potential for progressive improvement in athletic performance. If recovery from a training session takes longer for the MA, there is a greater risk of subsequent training bouts taking place before complete recovery has been achieved, leading to progressive decreases in performance potential, or *progressive overtraining* (Figure 6.2).

Alternatively, the MA would have a reduced capacity to train as hard at each session, or would have to prolong the recovery duration between training sessions. Either way, the outcome would be an impaired training response and a reduced effectiveness of the training program. At this point, it is tempting to attribute at least part of the progressive decline in performance for MAs to a longer recovery

81

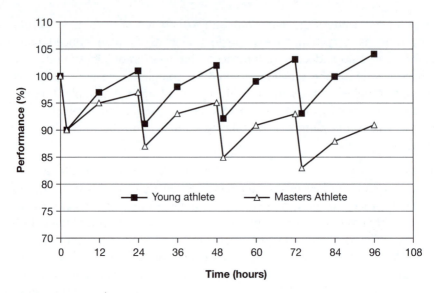

Figure 6.2 Theoretical performance response of two athletes undertaking the same training but demonstrating different recovery kinetics between training sessions (from Fell & Williams, 2008)

period that limits the ability to train effectively. For example, in an observational study, Pimentel et al. (2003) reported that, up until approximately 50 years of age, Masters runners maintained training volume and 10km running performance, but beyond this age, as training volume decreased, so to did 10km running performance. However, the evidence for impaired recovery in the MA is limited.

The age-related changes in both cardiac and skeletal muscle, the primary tissue type contributing to functional performance in the MA, have in general been investigated in non-athletic animals or humans, and this limits our ability to generalize much of this research to the training process for the MA. Nonetheless, the existing research may provide some clues to help explain the perception that recovery is impaired for the MA.

WHAT EVIDENCE EXISTS FOR IMPAIRED RECOVERY IN THE OLDER ATHLETE?

As skeletal muscle provides the contractile element that powers sport performance, any discussion of the effect of aging on recovery from exercise needs to address age-related alterations to this tissue. The following section will address the changes that have been observed in skeletal muscle with age, and then discuss the evidence that suggests these changes might contribute to impaired recovery in the aging athlete.

Age-related changes to skeletal muscle

Aging has been reported to result in reductions in the ability of the skeletal muscle to produce force (Frontera et al., 2000; Lexell et al., 1988; Prochniewicz et al., 2005). These changes appear to be related to decreases in both the volume of the muscle tissue (Frontera et al., 2000; Lexell et al., 1988) and a reduced ability to produce force per unit of volume (Frontera et al., 2000; Lowe et al., 2004; Metter et al., 1999). Both mechanisms for the reduction in function are addressed in more detail below.

Muscle size

There is a progressive loss of muscle fibers as part of the aging process. It has been reported that, by the eighth decade of life, the vastus lateralis muscles of elderly men contain on average 25 per cent fewer fibers than the corresponding muscles of men aged 18–37 years (Lexell et al., 1983), while by the ninth decade of life, this has decreased by 50 per cent (Lexell et al., 1988). More recent research using magnetic resonance imaging to describe changes in skeletal muscle mass with age indicates a reduction in total skeletal muscle mass by the end of the fifth decade (>45 years) that was more prevalent in the legs (30 per cent) than the arms (ten per cent) (Janssen et al., 2000).

In healthy young individuals, glycolytic muscle fibers have the largest diameter while oxidative fibers have the smallest (Hughes & Schiaffino, 1999). With increasing age, the muscles atrophy, with the glycolytic type II fibers decreasing in size (Frontera et al., 2000) such that in elderly subjects, the type IIX muscle fibers are the smallest, with the type I (Proctor et al., 1995) or type IIA fibers the largest (Essen-Gustavsson & Borges, 1986). Aging also results in shifts in muscle fiber proportions towards an increased percentage of the oxidative type I fibers at the expense of the glycolytic type IIX fibers (Aoyagi & Shephard, 1992), although it has been suggested that this phenomenon may not occur in healthy humans until the eighth decade of life (Frontera et al., 2000). There remains some contention as to how much the decreases in the relative area of the type II fibers are directly a consequence of aging, or whether they are

83

due to the reductions in physical activity that often accompany aging (Faulkner et al., 1995; Maharam et al., 1999). While it is possible that these changes may be the result of decreases in physical activity, increased relative area of type I fibers and decreased relative area of type IIB (IIX) fibers has recently been reported in old (53–77 years) compared to young (18–33 years) sprint-trained athletes (Korhonen et al., 2006).

Force per cross-sectional unit

In addition to the decline in muscle strength and function due to loss of muscle volume with age, there is evidence of declines in muscular strength that cannot be entirely explained by decreases in muscle volume (Akima et al., 2001; Metter et al., 1999). This qualitative decline in muscle strength with age is evident as a decrease in specific force (force per unit of muscle cross-sectional area) and has been attributed to factors such as the selective atrophy and overall decrease in the type II muscle fibers (Larsson et al., 1979), the denervation of motor units (Urbanchek et al., 2001), and to structural changes in the myosin molecules (Lowe et al., 2001). However, whether reductions in force per unit of muscle mass occur in Masters Athletes is uncertain, as many of the studies examining age-related reductions in force per unit of muscle mass have failed to consider physical activity levels.

In a recent study, Lowe et al. (2004) investigated the potential role of exercise training on age-related changes to myosin structure. In this experiment, old rats (~32 months) were exposed to five sets of six to ten maximal isometric contractions performed twice per week for four or eight to ten weeks via electrical stimulation. At the same time, a group of young adult (nine months) rats were subjected to reduced muscle activity via denervation of the semimembranosus muscle for either two or four weeks. Increased muscle activity in the old rats resulted in decreases in specific tension and fraction of myosin heads in the strong binding state after four weeks of training. After eight to ten weeks, however, specific tension had increased slightly from that of age-matched controls, and the fraction of myosin heads in the strong binding state had returned to normal. In contrast, decreased muscle activity in the young animals resulted in significant reductions in both specific tension and proportion of myosin heads in the strong binding state. The results of this study could be taken to support the contention that activity level rather than age contributes to the changes that have been observed in muscle quality. Alternatively, the short term decreases in specific tension might also be suggestive of an impaired adaption process for older muscle, which is the focus of a later section.

84

Changes in oxidative function

Mitochondria have a major role in cellular homeostasis, with particular emphasis on the bioenergetic status of the cell. Previously, healthy aging has been suggested to affect muscle oxidative capacity, with decreases in the activity of the oxidative enzyme citrate synthase reported for the gastrocnemius (Houmard et al., 1998) and the vastus lateralis (Essen-Gustavsson & Borges, 1986) muscles of elderly subjects. A more recent study by Hepple et al. (2003) reported that oxidative capacity of skeletal muscle is reduced in aged rats and that this is a major cause of age-related declines in $\dot{V}O_2$max. However, other studies in humans have reported no decline in the oxidative capacity of skeletal muscle as a result of aging using either in-vivo (Kent-Braun & Ng, 2000) or in-vitro (Rasmussen et al., 2003) techniques. Likely reasons for the conflicting findings that have been observed include the different techniques that have been used to measure oxidative function and the varying habitual activity levels of the older participants. The importance of activity level is demonstrated by the improvements in mitochondrial function that have been observed in older healthy (Berthon et al., 1995; Coggan et al., 1992; Frontera et al., 1990) and diseased (Williams et al., 2007) populations as a result of an exercise training intervention.

Despite the conflicting findings regarding alterations in the oxidative capacity of skeletal muscle with aging, recent studies have suggested that aging is associated with increasing numbers of dysfunctional mitochondria (Figueiredo et al., 2008a; Van Remmen & Richardson, 2001) and increased release of reactive oxygen species (ROS) (Conley et al., 2007a), both of which have been suggested to be responsible for a progressive loss of function in these organelles.

While the mitochondria have an important role in bioenergetics, they have also been identified as the primary cellular source of ROS (Stadtman, 2002). ROS are compounds which cause oxidative damage to the proteins, lipids, and DNA of the mitochondria (Figueiredo et al., 2008) and over time may lead to increasingly dysfunctional mitochondria, resulting in a diminished oxidative capacity of the cell and consequent loss of cellular function. Thus, ROS have been identified as likely contributors to the aging process (Stadtman, 2002). Relative levels of ROS production within the mitochondria is not, as is commonly believed, highest during high rates of oxygen consumption, but rather is greatest during periods of inactivity (Harper et al., 2004). Consequently, ROS production within the mitochondria may actually be exacerbated by decreases in physical activity.

85

Whether mitochondrial changes leading to increased ROS production is a problem for the MA remains equivocal. Waters et al. (2003) reported an age-related decline in mitochondrial function in healthy, physically active elderly; however this appeared to be somewhat attenuated by higher levels of physical activity, indicating that decrements in mitochondrial function may be due to reduced training stimulus with age. Several other studies have reported age-related declines in mitochondrial function even after accounting for physical activity levels (Conley et al., 2000; Tonkonogi et al., 2003), indicating that age-related deficits in mitochondrial function may not be solely due to physical deconditioning, but instead constitute an intrinsic element of biological aging. As mitogenesis is stimulated by physical activity (Adhihetty et al., 2003), it is possible that the maintenance of mitochondrial function observed with the more active participants in the study by Waters et al. (2003) was due to increased mitochondrial turnover. However, it is also possible that the findings may be due to a protective effect of the exercise training on existing mitochondria by upregulating antioxidant defense mechanisms (Vaanholt et al., 2008). If the increases in antioxidant defense and protein turnover mechanisms match the increase in ROS production with exercise training, the aging process may be ameliorated.

Although there is considerable evidence that the process of aging adversely affects skeletal muscle tissue in ways that may affect an athlete's ability to continue to generate high forces or to maintain moderate intensity contractions for pro-longed periods. The age at which this process commences is uncertain, with some age-related changes appearing to begin as early as the third decade of life (Inokuchi et al., 1975), while others may not commence until the eighth or ninth decade of life (Lexell et al., 1983; Lexell et al., 1988). Differences in the levels of physical activity undertaken by participants in many of the studies that have investigated the effect of aging on skeletal muscle are likely to be a major factor in the discordant results, and this may also apply to the following research that has addressed the effects of aging on recovery from exercise.

Fatigue and recovery research

There have been several comparisons of recovery from acute exercise between young and older subjects. Early research suggested that heart rate recovery after exercise was slowed with age (Cardus & Spencer, 1967). However, Darr et al. (1988) challenged this by attempting to match young and older groups for fitness level. Following incremental maximal exercise, trained participants demonstrated a faster heart rate recovery than untrained, but there were no age-related effects when young and old participants of the same fitness level were compared.

While heart rate recovery appears unaffected by age when physical fitness is considered, age might have a negative effect on acute recovery of skeletal muscle function after exercise. Klein et al. (1988) examined the contractile properties of the triceps surae muscle following electrically evoked muscle fatigue in a group of physically active young (19 to 32 years) and aging (64 to 69 years) men. Maximum voluntary contraction and vertical jump height were reduced similarly in both age groups following the fatiguing exercise, but one hour post-exercise, the rate of muscle relaxation after an electrically stimulated twitch contraction remained significantly reduced in the aging group compared to the young group. This slowed relaxation rate is suggestive of intracellular alterations leading to impairment of functional recovery. A possible explanation for this impaired function of the muscle may be related to a delayed metabolic recovery as a result of aging.

Several studies have used magnetic resonance spectroscopy (MRS) to investigate the effects of age on the recovery rate of muscle metabolites such as phosphocreatine (PCr) from exercise. Early MRS research (McCully et al., 1991) reported a slower time constant of PCr recovery (τPCr) with increased age after lower body (triceps surae) exercise, which was not improved by seven weeks of mild plantar flexion training in a healthy elderly group (80 $\pm$ 5 years). Similarly, in an investigation of the effects of creatine supplementation on quadriceps, PCr recovery rate from knee extension exercise in healthy young (31 $\pm$ 5 years) and older (58 $\pm$ 5 years) participants, Smith et al. (1998) reported that PCr resynthesis rate was initially (prior to oral creatine supplementation) lower in the older participants. In contrast, Kutsuzawa et al. (2001) found no differences in the τPCr between a group of young (28 $\pm$ 5 years) and older (61 $\pm$ 5 years) healthy sedentary subjects during recovery from three minutes of handgrip exercise. In a more recent MRS study using lower limb musculature (plantar flexion), there was no conclusive evidence that half recovery time for PCr was slower for older participants (73 $\pm$ 4 years) when matched with a younger group (25 $\pm$ 4 years) for activity level and body mass (Waters et al., 2003); however, the half recovery time for ADP levels to return to baseline was significantly longer in the older group irrespective of activity level. These conflicting findings make it difficult to clearly determine the impact of aging on acute muscle metabolite function, but indicate that differences in response by upper and lower muscle groups may exist, and highlight the need to control for activity level and body mass if investigating age-related changes in muscle metabolite recovery.

Of similar importance to muscle PCr levels, the resting (pre-exercise) concentration and the successful replenishment of muscle glycogen are also crucial for performance and recovery in athletes (Burke et al., 2001). Exercise can deplete muscle glycogen stores, which can negatively impact upon muscle function and athletic performance (Snyder, 1998). While the evidence suggests that aging muscle may have lower resting levels of both high energy phosphates (Moller et al., 1980; Tarnopolsky, 2000) and glycogen (Cartee, 1994), it is likely that this is largely due to a more sedentary lifestyle, as training has been shown to restore resting levels towards that of young muscle (Cartee, 1994; Meredith et al., 1989), and to maintain the rate at which muscle can take up glucose (Ivy et al., 1991). However, if aging was to cause a greater depletion during an exercise bout, or impair the restoration of muscle glycogen after exercise, then recovery may well be delayed for the MA.

Greater glycogen depletion seems an unlikely factor as studies that have used realistic exercise protocols at the same relative intensity have reported no difference in glycogen depletion between young and old muscle during the exercise bout (Cartee & Farrar, 1988; Nichols & Borer, 1987). However, aging has been shown to impair glucose tolerance and decrease insulin sensitivity (DeFronzo, 1981; Holloszy et al., 1986), which may affect glycogen restoration following exercise. Impaired glucose uptake may be due to age-related decreases in muscle glucose transporter (GLUT) levels, in particular GLUT-4 (Hall et al., 1994; Kern et al., 1992). However, in light of more recent research, this appears less likely for the healthy MA. When increases in body fatness are controlled for through caloric restriction, age-related differences in the action of insulin on skeletal muscle glycogen storage are no longer evident (Gupta et al., 2000). Furthermore, Cox et al. (1999) compared GLUT-4 concentration and insulin sensitivity in young and elderly men (20.9 ± 0.9 vs. 56.5 ± 1.9 yrs [± SE]) and women (22.4 ± 0.8 vs. 60.9 ± 1.0 yrs) both before and after seven days of endurance training and found no differences in GLUT-4 concentration before training, with an equivalent response to the training stimulus by both age groups despite higher body fatness and lower fitness levels in the older participants.

Despite the lack of evidence for an age-related impairment in glycogen replenishment in healthy MAs, there have been no studies to confirm this in response to typical training and competition loads. For the MA, an additional conundrum is that a combination of greater exercise-induced muscle damage and slower recovery may be synergistic in their impairment of muscle glycogen recovery. Exercise-induced muscle damage induces insulin resistance (Kirwan et al., 1992)

james fell and andrew williams

and impairs post-exercise muscle glycogen resynthesis (Costill et al., 1990) due to transient decreases in GLUT-4 protein content (Asp et al., 1995). If the MA was more susceptible to exercise-induced muscle damage, as has been suggested from studies of sedentary animals (Faulkner et al., 1990) and humans (Manfredi et al., 1991; Ploutz-Snyder et al., 2001), then this may also have a flow-on effect for recovery in terms of glycogen replenishment. Thus a combination of aging and damaging exercise may impair post-exercise glycogenesis to a greater extent than would be seen by either of these factors individually. Such a delayed recovery would be worse for the MA as this would decrease subsequent exercise capacity when performed prior to complete recovery of glycogen stores (Asp et al., 1998).

Evidence for greater muscle damage

Aging muscle is widely believed to be more susceptible to damage than younger muscle (McArdle et al., 2002), with two main mechanisms proposed. These include age-associated decreases in type II muscle fiber proportions (Frontera et al., 2000) and the possibility of increased production of ROS at rest and during exercise (Conley et al., 2007b).

As already addressed, aging results in decreases in the relative contributions of the high-force-producing type IIX muscle fibers to overall muscle mass (Aoyagi & Shephard, 1992), and in reduced numbers of muscle fibers (Inokuchi et al., 1975; Lexell et al., 1983). These ultra structural changes make it more likely that damage might occur due to the mechanical load exceeding the tensile strength of the tissue at any given intensity during exercise, or in the performance of normal daily activities. Certainly, the results of a number of studies support this as a potential mechanism for greater damage. Even if the rate of recovery is unaltered with aging, greater damage from any given mechanical load will require longer recovery periods and therefore reduce the number or intensity of training loads able to be undertaken by older athletes. Greater damage is also likely to increase pain sensation through an increased inflammatory response to the damage (Miles et al., 2008). However, for this to be a factor in increasing muscle damage in veteran athletes, there needs to have been ultra structural changes in the muscle. As previously discussed (Frontera et al., 2000; Janssen et al., 2000; Lexell et al., 1988), there is conflicting data in the literature about the age at which changes occur; consequently they may not be applicable to the majority of MAs.

The secondary route of increased or possibly accumulated damage in elderly muscle is via an increased production of ROS in the mitochondria both at rest and during exercise. This damage accumulates with age (Conley et al., 2007b) and can cause long-term harm to the mitochondria through mechanisms such as the oxidation of lipids along the inner mitochondrial membrane, resulting in alterations in structure and consequent increases in proton leak (Figueiredo et al., 2008b); or by damaging the mitochondrial DNA, resulting in mutations in this DNA and the resultant production of increasing numbers of dysfunctional mitochondria (Figueiredo et al., 2008b). In addition to the damage caused within the mitochondria, ROS that are released into or produced within the sarcoplasm can damage other molecules within the muscle cells (Figueiredo et al., 2008b). One possible site of damage is in the myosin heads, resulting in an altered structure and possibly impaired ability to form the strong binding state (Prochniewicz et al., 2005).

The body has several mechanisms to reduce or repair the damage caused by ROS. A defense system involving a range of antioxidant enzymes such as glutathione peroxidise, superoxide dismutase, and catalase, and nonenzymatic antioxidants including vitamins A, C, and E, exists to scavenge ROS before these molecules can cause damage to cellular structures. A second line of defense involves the repair or replacement of damaged macromolecules by processes such as protein turnover. While acute exercise results in increases in ROS production (Bailey et al., 2003), regular exercise training upregulates the activities of antioxidant enzymes (Jiménez-Jiménez et al., 2008) and has a stimulatory effect on protein synthesis rate in skeletal muscle (Vaanholt et al., 2008).

In a recent summation of their previous work investigating the effects of moderate exercise training on antioxidant function in rodents, Boveris and Navarro (2008) observed that mice subjected to treadmill exercise on a lifelong basis show a reduced mitochondrial content of thiobarbiturate reactive substances (TBARS) and protein carbonyls, suggesting reduced oxidative stress. However, these results were obtained from brain, liver, heart, and kidney tissue, which may respond to exercise differently from skeletal muscle.

Several studies have attempted to determine the role of physical activity on antioxidant defense systems and/or protein turnover in skeletal muscle from rodents with some reporting greater oxidative stress in the elderly animals (Gunduz et al., 2004; Vaanholt et al., 2008), while only a single study has reported reduced oxidative stress in skeletal muscle as a result of a lifetime of physical activity (Rosa et al., 2005). To date, no one has investigated the effect of long-term physical activity on antioxidant status in humans. The role that regular exercise training may play in protecting the skeletal muscle from damage, therefore, is uncertain.

Thus there are two potential causes for an increase in muscle damage with age. However, while the potential mechanisms of each cause have been described, the relative contribution of each to any age-associated increase in muscle damage is unclear. A confounding factor in considering these processes is that structural alterations that have been reported in the muscle with aging (decreased fiber numbers, smaller type II fibers) have been suggested to occur as a result of mitochondrial dysfunction (Conley et al., 2007b; Tarnopolsky & Safdar, 2008), which is likely due to accumulated oxidative damage. In addition, while these mechanisms have been described or inferred in multiple aging studies, little research has investigated mechanisms of damage with aging in physically active populations.

Evidence for impaired recovery

In addition to acute recovery of energy metabolite levels within the muscle after exercise, the preceding evidence indicates that with many types of vigorous training there may also be the added requirement for substantial tissue repair, which may be even greater in aging muscle. The ability for muscle tissue to recover from such damage is very important for restoration of pre-exercise function and the adaptation process (Figures 6.1 and 6.2). The concern for the MA is that the time taken for the muscle to repair and recover after fatiguing exercise or exercise-induced damage may be longer than for younger muscle.

Possible evidence for this concern comes from studies of rodents that have revealed significantly delayed repair and recovery from contraction-induced injury (Brooks & Faulkner, 1990; McBride et al., 1995; Rader & Faulkner, 2006a, 2006b; Zarzhevsky et al., 1999). A common approach has been to damage the muscle through lengthening contractions, and then compare the time course of recovery using measures such as force production or microscopic evidence of damage and repair.

Brooks and Faulkner (1990) used 15 minutes of lengthening contractions to induce similar reductions in force production (~34 per cent), and fiber number (~80 per cent) in young (two to three months) and older (26–27 months) mice. While the injured muscles of young mice had fully recovered by 28 days post-exercise, the muscles of the older mice remained incompletely recovered, and isometric force was still reduced at 60 days post-exercise. More recently, Rader and Faulkner compared the recovery of force after 225 electrically stimulated lengthening contractions in the plantar flexor muscles of adult (four to 13 months) and old (26–29 months) male (Rader & Faulkner, 2006b) and female (Rader & Faulkner, 2006a) mice. There were no differences in the extent of the injury between the two age groups in the first three days following the contraction protocol. However, while force deficits remained in both age groups at one month post-exercise, the young animals had recovered by two months, while the old mice showed incomplete recovery of isometric force and muscle mass, leading the authors to conclude that the changes in the old animals may have been permanent. This delayed recovery from lengthening contractions has also been reported by McBride et al. (1995), who found that the tibialis anterior of aging (32 months) rats took 14 days to return to pre-exercise functional levels compared with only five days for young adult (six months) muscle.

A delayed recovery of muscle function after damaging exercise has also been demonstrated in older humans (Dedrick & Clarkson, 1990; Klein et al., 1988). Dedrick and Clarkson (1990) found that lengthening contractions in college-aged women (23.6 ± 3.3 years) elicited the greatest strength loss immediately following the exercise, and thereafter strength demonstrated a progressive return to pre-exercise values after three days. In contrast, older women (67.4 ± 5.3 years) in the same study experienced a further decline in strength into the second day after exercise, and strength had not returned to pre-exercise levels after five days of recovery, remaining 38 per cent below pre-exercise values. Using a fatiguing protocol possibly more representative of exercise in the MA, Klein et al. (1988) reported a decrease in the rate of relaxation of twitch force and an increased half-relaxation time in the triceps surae muscle one hour after exercise in an older (64–69 years vs. 19–32 years) group of subjects. Furthermore, functional performance, as measured by maximal vertical jump height, remained significantly reduced (-9 per cent) after one hour of recovery in the older group only. This finding of a delayed recovery of dynamic muscle function after fatiguing exercise suggests that recovery of power may take longer for the MA and could well have a negative impact on functional sporting capability after exercise-induced fatigue.

Recovery of maximal contraction force may be better maintained with age if the exercise is fatiguing but not specifically designed to elicit damage. In the above study by Klein et al. (1988), there were no differences between the younger and older groups in restoration of muscle force during the one hour of recovery after fatigue. Similarly, following fatiguing isometric contractions, the recovery profiles for force, contractile speed, surface electromyography, muscle activation via twitch interpolation, and muscle compound action potentials in the elbow flexors of young (24 ± 2 years) and older (84 ± 2 years) men, were not different when parameters were normalized to the pre-fatigue value (Allman & Rice, 2001). Even damaging exercise may not lead to a delayed recovery time when there are no initial differences in maximal isometric force between the age groups investigated. Lavender and Nosaka (2008) recently compared the loss and recovery of maximal isometric force after lengthening contractions of the elbow flexors in young (19.4 ± 0.4 years) and middle-aged (48.0 ± 2.1

years) untrained men and found that there were no differences in the time course of muscle force or recovery between the two groups.

To date, only two studies have compared recovery from exercise in trained young and older participants. McLester et al. (2003) investigated functional recovery 24, 48, 72, and 96 hours after an acute bout of resistance exercise in resistance-trained subjects. A significant difference was observed between younger (18–30 years) and older (50–65 years) subjects in the number of repetitions performed at 72 hours post exercise. In contrast, Fell et al. (2006) did not find any differences in functional recovery between young (24 ± 5 years) and Masters cyclists (45 ± 6 years) in response to three days of a repeated endurance task (30-minute cycling time trial). The difference between these studies may be due to the type of fatiguing exercise employed (resistance compared with endurance exercise) and highlight the need for further research with this population.

AGING AND ADAPTIVE POTENTIAL

The research evidence that has demonstrated greater fatigue or damage and a slower rate of recovery from exercise as a consequence of aging provides evidence that these factors might negatively influence the adaptation process. Aging may retard the long-term response to a training stimulus, thus reducing or negating the potential gains in athletic performance, a major goal for many MAs. Research on animals demonstrating how older muscle may suffer permanent reductions in functional capacity after damaging exercise or immobilization (Rader & Faulkner, 2006a; Zarzhevsky et al., 2001) provides evidence for such an argument. However, there is also substantial evidence that regular exercise provides protection against many physiological and functional declines associated with aging (McArdle et al., 2002; Radak et al., 2001; Vincent et al., 2002; Wang et al., 2002). For the MA, the ability of skeletal muscle to respond and adapt to exercise overload is fundamental to the training process.

Several variables involved in physiological adaptation processes have been proposed to be affected by aging. Inflammatory (Toft et al., 2002), genetic (Jozsi et al., 2000, 2001), hormonal (Kraemer et al., 1999), and satellite cell (Carlson, 1995; Grounds, 1998) responses have all been investigated and proposed as mechanisms for impaired adaptation potential in aging muscle. However, in human studies where training responses for different-aged subjects to a given exercise stimulus are compared, the results are still equivocal.

Early work by McBride et al. (1995) found that the repeated-bout effect (protection from damaging exercise after a prior exposure to damaging exercise) was impaired in aged rat muscle (32 months) in comparison with adult muscle (six months). However, more recent research in rodents has demonstrated that even non-damaging exercise such as isometric contractions or passive stretches can provide a protective effect from a subsequent bout of potentially damaging exercise (lengthening contractions) equally in both young (three months) and older (24 months) mice (Koh et al., 2003). In humans, there are conflicting findings regarding the repeated-bout effect. Early work reported that older women demonstrated the same ability to adapt to damaging exercise as young women when exposed to a second damaging exercise bout seven days later (Clarkson & Dedrick, 1988). In contrast, more recent research has proposed that for older men (70.5 ± 4.1 years) the protective effect conferred by an initial bout of damaging exercise is less than in younger men (Lavender & Nosaka, 2006). The reasons for these discordant findings may be due to gender differences or that the latter study did not perform the repeated bout until four weeks after the first, which may be an indication that training adaptations are lost faster in older muscle.

A positive finding for the MA is that regular resistance training has been shown to reduce susceptibility to exercise-induced muscle damage in older women (66 ± 5 years) to the same level as younger (23 ± 4 years) women (Ploutz-Snyder et al., 2001). Furthermore, comparisons between younger and older humans for improvement in various physiological and functional measures such as muscle hypertrophy (Ivey et al., 2000) and cardiovascular fitness (Kohrt et al., 1991) have in general found no differences, suggesting that the plasticity of skeletal muscle is retained with age (Galvao et al., 2005). While it is inevitable that aging will eventually lead to the deterioration of many physiological variables, many of the studies to examine these variables in muscle tissue have used exercise protocols that are incongruent with normal exercise training. Moreover, well-structured training programs appear likely to elicit effective responses regardless of age.

LIMITATIONS AND FUTURE DIRECTIONS

Regardless of the weight of evidence for or against delayed recovery in the aging athlete, it is clear that many MAs believe that delayed recovery is a genuine issue (Reaburn, 2004), and report that recovery is impaired despite the absence of any measurable decrease in performance (Fell et al., 2008). Interpreting such anecdotal evidence for an impaired recovery with very little empirical evidence for such a phenomenon leaves the MA in somewhat of a conundrum. Best practice in structuring training for athletes of any age requires careful consideration of the principles of fatigue and recovery. Adequate hydration and optimal nutrition, appropriate periodization of the training plan,

94

and the potential use of physical therapy and supplementation post training and competition, should all be given the fullest attention by athletes and coaches. Regular monitoring of training load can be achieved in a number of ways, enabling the MA to respond appropriately to any abnormal reactions to training load. If, indeed, recovery is delayed with aging, maintaining intensity of training sessions but reducing volume may be a practical solution, given that most MAs probably already have a long history of high training volumes. Continued attention to core principles of training and further research into the physiology and psychology of these unique athletes will hopefully enable a continuation of these trends.

This chapter has presented a selection of studies that have examined the effect of aging on muscle recovery from exercise. The research presented has provided conflicting evidence with respect to the effect of age on muscle recovery, repair, and adaptation after exercise. However, difficulties exist in the comparison of many of these studies due to methodological differences that may well account for the conflicting findings. Of importance when considering these methodological differences are factors such as the type of exercise protocol used to elicit fatigue and the different relative ages of the research subjects from adult and middle-aged to very old and senescent. Considering that several studies have identified that there may be critical ages to which muscle function can be maintained if undertaking appropriate training protocols, but beyond which, rapid decrements in function are unavoidable (Galloway et al., 2002; Pimentel et al., 2003), clearer age-group classification guidelines may be required. Finally, the level of habitual activity or training performed by the participants in many studies that consider aging and muscle function is also likely to contribute to the discordant findings reported in the literature. Most studies have examined participants or animals from completely sedentary lifestyles, while some have used human participants that are active, but rarely athletic. Consequently, at present there are many difficulties in translating research findings on the indices of muscle recovery in very old, inactive rats after severe lengthening contractions, with recovery from training-induced fatigue in the MA.

In order for any future research to more clearly define the true effect of aging on recovery in the MA, it should control for the training status of participants and incorporate exercise models that are representative of the type of training regularly undertaken by participants. Only then can the effect of aging on exercise-induced muscle damage and functional recovery be genuinely compared without the confounding influence of age-related declines in training load.

Postnote: Dara Torres enjoyed a successful Olympic campaign in 2008, winning silver medals in the 50m freestyle, 4x100m freestyle relay, and 4x100m medley relay.

REFERENCES

Adhihetty, P.J., Irrcher, I., Joseph, A.M., Ljubicic, V., & Hood, D.A. (2003). Plasticity of skeletal muscle mitochondria in response to contractile activity. *Experimental Physiology*, 88, 99–107.

Akima, H., Kano, Y., Enomoto, Y., Ishizu, M., Okada, M., Oishi, Y., Katsuta, S., & Kuno, S. (2001). Muscle function in 164 men and women aged 20–84 yr. *Medicine and Science in Sports and Exercise*, 33, 220–226.

Allman, B.L., & Rice, C.L. (2001). Incomplete recovery of voluntary isometric force after fatigue is not affected by old age. *Muscle and Nerve*, 24, 1156–1167.

Aoyagi, Y., & Shephard, R.J. (1992). Aging and muscle function. *Sports Medicine*, 14, 376–396.

Asp, S., Kristiansen, S., & Richter, E.A. (1995). Eccentric muscle damage transiently decreases rat skeletal muscle GLUT-4 protein. *Journal of Applied Physiology*, 79, 1338–1345.

Asp, S., Daugaard, J.R., Kristiansen, S., Kiens, B., & Richter, E.A. (1998). Exercise metabolism in human skeletal muscle exposed to prior eccentric exercise. *Journal of Physiology*, 509, 305–313.

Bailey, D.M., Davies, B., Young, I.S., Jackson, M.J., Davison, G.W., Isaacson, R., & Richardson, R.S. (2003). EPR spectroscopic detection of free radical outflow from an isolated muscle bed in exercising humans. *Journal of Applied Physiology*, 94, 1714–1718.

Berthon, P., Freyssenet, D., Chatard, J.C., Castells, J., Mujika, I., Geyssant, A., Guezennec, C.Y., & Dennis, C. (1995). Mitochondrial ATP production rate in 55- to 73-year-old men: effect of endurance training. *Acta Physiologica Scandinavica*, 154, 269–274.

Bompa, T.O. (1999). Rest and recovery. In T.O. Bompa (Ed.), *Periodization: Theory and Methodology of Training* (4th ed., pp. 95–142). Champaign, IL: Human Kinetics.

Boveris, A., & Navarro, A. (2008). Systemic and mitochondrial adaptive responses to moderate exercise in rodents. *Free Radical Biology and Medicine*, 44, 224–229.

Brooks, S.V., & Faulkner, J.A. (1990). Contraction-induced injury: recovery of skeletal muscles in young and old mice. *American Journal of Physiology*, 258, C436–442.

Burke, L.M., Cox, G.R., Cummings, N.K., & Desbrow, B. (2001). Guidelines for daily carbohydrate intake: do athletes achieve them? *Sports Medicine*, 31, 267–299.

Cardus, D., & Spencer, W.A. (1967). Recovery time of heart frequency in healthy men: its relation to age and physical condition. *Archives of Physical Medicine and Rehabilitation*, 48, 71–77.

Carlson, B.M. (1995). Factors influencing the repair and adaptation of muscles in aged individuals: satellite cells and innervation. *Journals of Gerontology. Series A, Biological Sciences and Medical Sciences*, 50A Spec, 96–100.

Cartee, G.D. (1994). Aging skeletal muscle: response to exercise. *Exercise and Sport Sciences Reviews*, 22, 91–120.

Cartee, G.D., & Farrar, R.P. (1988). Exercise training induces glycogen sparing during exercise by old rats. *Journal of Applied Physiology*, 64, 259–265.

Clarkson, P.M., & Dedrick, M.E. (1988). Exercise-induced muscle damage, repair, and adaptation in old and young subjects. *Journal of Gerontology*, 43, M91–M96.

96

Coggan, A.R., Spina, R.J., King, D.S., Rogers, M.A., Brown, M., Nemeth, P.M., & Holloszy, J.O. (1992). Skeletal muscle adaptations to endurance training in 60- to 70-yr-old men and women. *Journal of Applied Physiology, 72*, 1780–1786.

Conley, K.E., Jubrias, S.A., & Esselman, P.C. (2000). Oxidative capacity and ageing in human muscle. *Journal of Physiology, 526*, 203–210.

Conley, K.E., Marcinek, D.J., & Villarin, J. (2007a). Mitochondrial dysfunction and age. *Current Opinion in Clinical Nutrition and Metabolic Care, 10*, 688–692.

Conley, K.E., Amara, C.E., Jubrias, S.A., & Marcinek, D.J. (2007b). Mitochondrial function, fibre types and ageing: new insights from human muscle in vivo. *Experimental Physiology, 92*, 333–339.

Costill, D.L., Pascoe, D.D., Fink, W.J., Robergs, R.A., Barr, S.I., & Pearson, D. (1990). Impaired muscle glycogen resynthesis after eccentric exercise. *Journal of Applied Physiology, 69*, 46–50.

Cox, J.H., Cortright, R.N., Dohm, G.L., & Houmard, J.A. (1999). Effect of aging on response to exercise training in humans: skeletal muscle GLUT-4 and insulin sensitivity. *Journal of Applied Physiology, 86*, 2019–2025.

Cuda, A. (2008, 9th July 2008). Athletes compete into later years. Connecticut Post. Retrieved 10th July 2008 from http://www.connpost.com/localnews/ci_9832994

Darr, K.C., Bassett, D.R., Morgan, B.J., & Thomas, D.P. (1988). Effects of age and training status on heart rate recovery after peak exercise. *American Journal of Physiology, 254*, H340–343.

Dedrick, M.E., & Clarkson, P.M. (1990). The effects of eccentric exercise on motor performance in young and older women. *European Journal of Applied Physiology and Occupational Physiology, 60*, 183–186.

DeFronzo, R.A. (1981). Glucose intolerance and aging. *Diabetes Care, 4*, 493–501.

Essen-Gustavsson, B., & Borges, O. (1986). Histochemical and metabolic characteristics of human skeletal muscle in relation to age. *Acta Physiologica Scandinavica, 126*, 107–114.

Faulkner, J.A., Brooks, S.V., & Zerba, E. (1990). Skeletal muscle weakness and fatigue in old age: underlying mechanisms. *Annual Review of Gerontology and Geriatrics, 10*, 147–166.

Faulkner, J.A., Brooks, S.V., & Zerba, E. (1995). Muscle atrophy and weakness with aging: contraction-induced injury as an underlying mechanism. *Journals of Gerontology. Series A, Biological Sciences and Medical Sciences, 50A Spec*, 124–129.

Fell, J., Haseler, L., Gaffney, P., Reaburn, P., & Harrison, G. (2006). Performance during consecutive days of laboratory time-trials in young and veteran cyclists. *Journal of Sports Medicine and Physical Fitness, 46*, 395–402.

Fell, J., Reaburn, P., & Harrison, G.J. (2008). Altered perception and report of fatigue and recovery in veteran athletes. *Journal of Sports Medicine and Physical Fitness, 48*, 272–276.

Fell, J., & Williams, A. D. (2008). The effect of aging on skeletal-muscle recovery from exercise: Possible implications for aging athletes. *Journal of Aging and Physical Activity, 16*, 97–115.

Figueiredo, P.A., Ferreira, R.M., Appell, H.J., & Duarte, J.A. (2008a). Age-induced morphological, biochemical, and functional alterations in isolated mitochondria from murine skeletal muscle. *Journals of Gerontology. Series A, Biological Sciences and Medical Sciences, 63*, 350–359.

97

Figueiredo, P.A., Mota, M.P., Appell, H.J., & Duarte, J.A. (2008b). The role of mitochondria in aging of skeletal muscle. *Biogerontology*, 9, 67–84.

Frontera, W.R., Meredith, C.N., O'Reilly, K.P., & Evans, W.J. (1990). Strength training and determinants of $\dot{V}O_2$max in older men. *Journal of Applied Physiology*, 68, 329–333.

Frontera, W.R., Hughes, V.A., Fielding, R.A., Fiatarone, M.A., Evans, W.J., & Roubenoff, R. (2000). Aging of skeletal muscle: a 12-yr longitudinal study. *Journal of Applied Physiology*, 88, 1321–1326.

Galloway, M.T., Kadoko, R., & Jokl, P. (2002). Effect of aging on male and female master athletes' performance in strength versus endurance activities. *American Journal of Orthopedics*, 31, 93–98.

Galvao, D.A., Newton, R.U., & Taaffe, D.R. (2005). Anabolic responses to resistance training in older men and women: a brief review. *Journal of Aging and Physical Activity*, 13, 343–358.

Grounds, M.D. (1998). Age-associated changes in the response of skeletal muscle cells to exercise and regeneration. *Annals of the New York Academy of Sciences*, 854, 78–91.

Gunduz, F., Senturk, U.K., Kuru, O., Aktekin, B., & Aktekin, M.R. (2004). The effect of one year's swimming exercise on oxidant stress and antioxidant capacity in aged rats. *Physiological Research*, 53, 171–176.

Gupta, G., She, L., Ma, X.-H., Yang, X.-M., Hu, M., Cases, J.A., Vuguin, P., Rossetti, L., & Barzilai, N. (2000). Aging does not contribute to the decline in insulin action on storage of muscle glycogen in rats. *American Journal of Physiology-Regulatory Integrative and Comparative Physiology*, 278, R111–117.

Hall, J.L., Mazzeo, R.S., Podolin, D.A., Cartee, G.D., & Stanley, W.C. (1994). Exercise training does not compensate for age-related decrease in myocardial GLUT-4 content. *Journal of Applied Physiology*, 76, 328–332.

Harper, M.E., Bevilacqua, L., Hagopian, K., Weindruch, R., & Ramsey, J.J. (2004). Ageing, oxidative stress, and mitochondrial uncoupling. *Acta Physiologica Scandinavica*, 182, 321–331.

Hepple, R.T., Hagen, J.L., Krause, D.J., & Jackson, C.C. (2003). Aerobic power declines with aging in rat skeletal muscles perfused at matched convective O_2 delivery. *Journal of Applied Physiology*, 94, 744–751.

Holloszy, J.O., Schultz, J., Kusnierkiewicz, J., Hagberg, J.M., & Ehsani, A.A. (1986). Effects of exercise on glucose tolerance and insulin resistance. Brief review and some preliminary results. *Acta Medica Scandinavica. Supplementum*, 711, 55–65.

Houmard, J.A., Weidner, M.L., Gavigan, K.E., Tyndall, G.L., Hickey, M.S., & Alshami, A. (1998). Fiber type and citrate synthase activity in the human gastrocnemius and vastus lateralis with aging. *Journal of Applied Physiology*, 85, 1337–1341.

Hughes, S.M., & Schiaffino, S. (1999). Control of muscle fibre size: a crucial factor in ageing. *Acta Physiologica Scandanavica*, 167, 307–312.

Inokuchi, S., Ishikawa, H., Iwamoto, S., & Kimura, T. (1975). Age-related changes in the histological composition of the rectus abdominis muscle of the adult human. *Human Biology*, 47, 231–249.

Ivey, F.M., Roth, S.M., Ferrell, R.E., Tracy, B.L., Lemmer, J.T., Hurlbut, D.E., Martel, G.F., Siegel, E.F., Fozard, J.L., Jeffrey Metter, E., Fleg, J.L., & Hurley, B.F. (2000). Effects of age, gender, and myostatin genotype on the hypertrophic response to heavy resistance strength training. *The Journal of Gerontology Series A, Biological Sciences and Medical Sciences*, 55, M641–648.

Ivy, J.L., Young, J.C., Craig, B.W., Kohrt, W.M., & Holloszy, J.O. (1991). Ageing, exercise and food restriction: effects on skeletal muscle glucose uptake. *Mechanisms of Ageing and Development*, 61, 123–133.

Janssen, I., Heymsfield, S.B., Wang, Z.M., & Ross, R. (2000). Skeletal muscle mass and distribution in 468 men and women aged 18–88 yr. *Journal of Applied Physiology*, 89, 81–88.

Jiménez-Jiménez, R., Cuevas, M.J., Almar, M., Lima, E., García-López, D., De Paz, J.A., & González-Gallego, J. (2008). Eccentric training impairs NF-[kappa]B activation and over-expression of inflammation-related genes induced by acute eccentric exercise in the elderly. *Mechanisms of Ageing and Development*, 129, 313–321.

Jozsi, A.C., Dupont-Versteegden, E.E., Taylor-Jones, J.M., Evans, W.J., Trappe, T.A., Campbell, W.W., & Peterson, C.A. (2000). Aged human muscle demonstrates an altered gene expression profile consistent with an impaired response to exercise. *Mechanisms of Ageing and Development*, 120, 45–56.

Jozsi, A.C., Dupont-Versteegden, E.E., Taylor-Jones, J.M., Evans, W.J., Trappe, T.A., Campbell, W.W., & Peterson, C.A. (2001). Molecular characteristics of aged muscle reflect an altered ability to respond to exercise. *International Journal of Sport Nutrition and Exercise Metabolism*, 11 (Suppl.), S9–S15.

Kent-Braun, J.A., & Ng, A.V. (2000). Skeletal muscle oxidative capacity in young and older women and men. *Journal of Applied Physiology*, 89, 1072–1078.

Kern, M., Dolan, P.L., Mazzeo, R.S., Wells, J.A., & Dohm, G.L. (1992). Effect of aging and exercise on GLUT-4 glucose transporters in muscle. *American Journal of Physiology-Endocrinology and Metabolism*, 263, E362–367.

Kirwan, J.P., Hickner, R.C., Yarasheski, K.E., Kohrt, W.M., Wiethop, B.V., & Holloszy, J.O. (1992). Eccentric exercise induces transient insulin resistance in healthy individuals. *Journal of Applied Physiology*, 72, 2197–2202.

Klein, C., Cunningham, D.A., Paterson, D.H., & Taylor, A.W. (1988). Fatigue and recovery contractile properties of young and elderly men. *European Journal of Applied Physiology and Occupational Physiology*, 57, 684–690.

Koh, T.J., Peterson, J.M., Pizza, F.X., & Brooks, S.V. (2003). Passive stretches protect skeletal muscle of adult and old mice from lengthening contraction-induced injury. *Journals of Gerontology. Series A, Biological Sciences and Medical Sciences*, 58, 592–597.

Kohrt, W.M., Malley, M.T., Coggan, A.R., Spina, R.J., Ogawa, T., Ehsani, A.A., Bourey, R.E., Martin, W.H., III, & Holloszy, J.O. (1991). Effects of gender, age, and fitness level on response of $\dot{V}O_2$max to training in 60–71 yr olds. *Journal of Applied Physiology*, 71, 2004–2011.

Korhonen, M.T., Cristea, A., Alen, M., Hakkinen, K., Sipila, S., Mero, A., Viitasalo, J.T., Larsson, L., & Suominen, H. (2006). Aging, muscle fiber type, and contractile function in sprint-trained athletes. *Journal of Applied Physiology*, 101, 906–917.

Koutedakis, Y., Metsios, G.S., & Stavropoulos-Kalinoglou, A. (2006). Periodization of exercise training in sport. In G. Whyte (Ed.), *The Physiology of Training*. Philadelphia: Elsevier.

Kraemer, W.J., Hakkinen, K., Newton, R.U., Nindl, B.C., Volek, J.S., McCormick, M., Gotshalk, L.A., Gordon, S.E., Fleck, S.J., Campbell, W.W., Putukian, M.,

& Evans, W.J. (1999). Effects of heavy-resistance training on hormonal response patterns in younger vs. older men. *Journal of Applied Physiology, 87*, 982–992.

Kutsuzawa, T., Shioya, S., Kurita, D., Haida, M., & Yamabayashi, H. (2001). Effects of age on muscle energy metabolism and oxygenation in the forearm muscles. *Medicine and Science in Sports and Exercise, 33*, 901–906.

Larsson, L., Grimby, G., & Karlsson, J. (1979). Muscle strength and speed of movement in relation to age and muscle morphology. *Journal of Applied Physiology, 46*, 451–456.

Lavender, A.P., & Nosaka, K. (2006). Responses of old men to repeated bouts of eccentric exercise of the elbow flexors in comparison with young men. *European Journal of Applied Physiology, 97*, 619–626.

Lavender, A.P., & Nosaka, K. (2008). Changes in markers of muscle damage of middle-aged and young men following eccentric exercise of the elbow flexors. *Journal of Science and Medicine in Sport, 11*, 124–131.

Lexell, J., Henriksson-Larsen, K., Winblad, B., & Sjostrom, M. (1983). Distribution of different fiber types in human skeletal muscles: effects of aging studied in whole muscle cross sections. *Muscle and Nerve, 6*, 588–595.

Lexell, J., Taylor, C.C., & Sjostrom, M. (1988). What is the cause of the ageing atrophy? Total number, size and proportion of different fiber types studied in whole vastus lateralis muscle from 15- to 83-year-old men. *Journal of the Neurological Sciences, 84*, 275–294.

Lowe, D.A., Surek, J.T., Thomas, D.D., & Thompson, L.V. (2001). Electron paramagnetic resonance reveals age-related myosin structural changes in rat skeletal muscle fibers. *American Journal of Physiology-Cell Physiology, 280*, C540–547.

Lowe, D.A., Husom, A.D., Ferrington, D.A., & Thompson, L.V. (2004). Myofibrillar myosin ATPase activity in hindlimb muscles from young and aged rats. *Mechanisms of Ageing and Development, 125*, 619–627.

Maharam, L.G., Bauman, P.A., Kalman, D., Skolnik, H., & Perle, S.M. (1999). Masters athletes: factors affecting performance. *Sports Medicine, 28*, 273–285.

Manfredi, T.G., Fielding, R.A., O'Reilly, K.P., Meredith, C.N., Lee, H.Y., & Evans, W.J. (1991). Plasma creatine kinase activity and exercise-induced muscle damage in older men. *Medicine and Science in Sports and Exercise, 23*, 1028–1034.

McArdle, A., Vasilaki, A., & Jackson, M. (2002). Exercise and skeletal muscle ageing: cellular and molecular mechanisms. *Ageing Research Reviews, 1*, 79–93.

McBride, T.A., Gorin, F.A., & Carlsen, R.C. (1995). Prolonged recovery and reduced adaptation in aged rat muscle following eccentric exercise. *Mechanisms of Ageing and Development, 83*, 185–200.

McCully, K.K., Forciea, M.A., Hack, L.M., Donlon, E., Wheatley, R.W., Oatis, C.A., Goldberg, T., & Chance, B. (1991). Muscle metabolism in older subjects using 31P magnetic resonance spectroscopy. *Canadian Journal of Physiology and Pharmacology, 69*, 576–580.

McLester, J.R., Bishop, P.A., Smith, J., Wyers, L., Dale, B., Kozusko, J., Richardson, M., Nevett, M.E., & Lomax, R. (2003). A series of studies-a practical protocol for testing muscular endurance recovery. *Journal of Strength and Conditioning Research, 17*, 259–273.

Meredith, C.N., Frontera, W.R., Fisher, E.C., Hughes, V.A., Herland, J.C., Edwards, J., & Evans, W.J. (1989). Peripheral effects of endurance training in young and old subjects. *Journal of Applied Physiology, 66*, 2844–2849.

Metter, E.J., Lynch, N., Conwit, R., Lindle, R., Tobin, J., & Hurley, B. (1999). Muscle quality and age: cross-sectional and longitudinal comparisons. *Journals of Gerontology. Series A, Biological Sciences and Medical Sciences*, 54, B207–218.

Miles, M.P., Andring, J.M., Pearson, S.D., Gordon, L.K., Kasper, C., Depner, C.M., & Kidd., J.R. (2008). Diurnal variation, response to eccentric exercise, and association of inflammatory mediators with muscle damage variables. *Journal of Applied Physiology*, 104, 451–458.

Moller, P., Bergstrom, J., Furst, P., & Hellstrom, K. (1980). Effect of aging on energy-rich phosphagens in human skeletal muscles. *Clinical Science*, 58, 553–555.

Nichols, J.F., & Borer, K.T. (1987). The effects of age on substrate depletion and hormonal responses during submaximal exercise in hamsters. *Physiology and Behavior*, 41, 1–6.

Pimentel, A.E., Gentile, C.L., Tanaka, H., Seals, D.R., & Gates, P.E. (2003). Greater rate of decline in maximal aerobic capacity with age in endurance-trained than in sedentary men. *Journal of Applied Physiology*, 94, 2406–2413.

Ploutz-Snyder, L.L., Giamis, E.L., Formikell, M., & Rosenbaum, A.E. (2001). Resistance training reduces susceptibility to eccentric exercise-induced muscle dysfunction in older women. *Journals of Gerontology. Series A, Biological Sciences and Medical Sciences*, 56, B384–B390.

Prochniewicz, E., Thomas, D.D., & Thompson, L.V. (2005). Age-related decline in actomyosin function. *Journals of Gerontology. Series A, Biological Sciences and Medical Sciences*, 60, 425–431.

Proctor, D.N., Sinning, W.E., Walro, J.M., Sieck, G.C., & Lemon, P.W. (1995). Oxidative capacity of human muscle fiber types: effects of age and training status. *Journal of Applied Physiology*, 78, 2033–2038.

Radak, Z., Kaneko, T., Tahara, S., Nakamoto, H., Pucsok, J., Sasvári, M., Nyakas, C., & Goto, S. (2001). Regular exercise improves cognitive function and decreases oxidative damage in rat brain. *Neurochemistry International*, 38, 17–23.

Rader, E.P., & Faulkner, J.A. (2006a). Effect of aging on the recovery following contraction-induced injury in muscles of female mice. *Journal of Applied Physiology*, 101, 887–892.

Rader, E.P., & Faulkner, J.A. (2006b). Recovery from contraction-induced injury is impaired in weight-bearing muscles of old male mice. *Journal of Applied Physiology*, 100, 656–661.

Rasmussen, U.F., Krustrup, P., Kjaer, M., & Rasmussen, H.N. (2003). Human skeletal muscle mitochondrial metabolism in youth and senescence: no signs of functional changes in ATP formation and mitochondrial oxidative capacity. Pflugers Archiv. *European Journal of Physiology*, 446, 270–278.

Reaburn, P. (2004). Recovery for ageing athletes. *Sports Coach*, 26, 12–14.

Rosa, E.F., Silva, A.C., Ihara, S.S., Mora, O.A., Aboulafia, J., & Nouailhetas, V.L. (2005). Habitual exercise program protects murine intestinal, skeletal, and cardiac muscles against aging. *Journal of Applied Physiology*, 99, 569–575.

Smith, J.C., Stephens, D.P., Hall, E.L., Jackson, A.W., & Earnest, C.P. (1998). Effect of oral creatine ingestion on parameters of the work rate-time relationship and time to exhaustion in high-intensity cycling. *European Journal of Applied Physiology and Occupational Physiology*, 77, 360–365.

Snyder, A.C. (1998). Overtraining and glycogen depletion hypothesis. *Medicine and Science in Sports and Exercise*, 30, 1146–1150.

Stadtman, E.R. (2002). Importance of individuality in oxidative stress and aging. *Free Radical Biology and Medicine*, 33, 597–604.

Tarnopolsky, M.A. (2000). Potential benefits of creatine monohydrate supplementation in the elderly. *Current Opinion in Clinical Nutrition and Metabolic Care*, 3, 497–502.

Tarnopolsky, M.A., & Safdar, A. (2008). The potential benefits of creatine and conjugated linoleic acid as adjuncts to resistance training in older adults. *Applied Physiology, Nutrition, and Metabolism*, 33, 213–227.

Toft, A.D., Jensen, L.B., Bruunsgaard, H., Ibfelt, T., Halkjaer-Kristensen, J., Febbraio, M., & Pedersen, B.K. (2002). Cytokine response to eccentric exercise in young and elderly humans. *American Journal of Physiology-Cell Physiology*, 283, C289–C295.

Tonkonogi, M., Fernstrom, M., Walsh, B., Ji, L.L., Rooyackers, O., Hammarqvist, F., Wernerman, J., & Sahlin, K. (2003). Reduced oxidative power but unchanged antioxidative capacity in skeletal muscle from aged humans. Pflugers Archiv. *European Journal of Physiology*, 446, 261–269.

Urbanchek, M.G., Picken, E.B., Kalliainen, L.K., & Kuzon, W.M., Jr. (2001). Specific force deficit in skeletal muscles of old rats is partially explained by the existence of denervated muscle fibers. *Journals of Gerontology. Series A, Biological Sciences and Medical Sciences*, 56, B191–197.

Vaanholt, L.M., Speakman, J.R., Garland, T., Jr., Lobley, G.E., & Visser, G.H. (2008). Protein synthesis and antioxidant capacity in aging mice: effects of long-term voluntary exercise. *Physiological and Biochemical Zoology*, 81, 148–157.

Van Remmen, H., & Richardson, A. (2001). Oxidative damage to mitochondria and aging. *Experimental Gerontology*, 36, 957–968.

Vincent, K.R., Vincent, H.K., Braith, R.W., Lennon, S.L., & Lowenthal, D.T. (2002). Resistance exercise training attenuates exercise-induced lipid peroxidation in the elderly. *European Journal of Applied Physiology*, 87, 416–423.

Wang, B.W., Ramey, D.R., Schettler, J.D., Hubert, H.B., & Fries, J.F. (2002). Postponed development of disability in elderly runners: a 13-year longitudinal study. *Archives of Internal Medicine*, 162, 2285–2294.

Waters, D.L., Brooks, W.M., Qualls, C.R., & Baumgartner, R.N. (2003). Skeletal muscle mitochondrial function and lean body mass in healthy exercising elderly. *Mechanisms of Ageing and Development*, 124, 301–309.

Williams, A.D., Carey, M.F., Selig, S., Hayes, A., Krum, H., Patterson, J., Toia, D., & Hare, D.L. (2007). Circuit resistance training in chronic heart failure improves skeletal muscle mitochondrial ATP production rate-a randomized controlled trial. *Journal of Cardiac Failure*, 13, 79–85.

Zarzhevsky, N., Carmeli, E., Fuchs, D., Coleman, R., Stein, H., & Reznick, A.Z. (2001). Recovery of muscles of old rats after hindlimb immobilisation by external fixation is impaired compared with those of young rats. *Experimental Gerontology*, 36, 125–140.

Zarzhevsky, N., Coleman, R., Volpin, G., Fuchs, D., Stein, H., & Reznick, A.Z. (1999). Muscle recovery after immobilisation by external fixation. *Journal of Bone and Joint Surgery. British Volume*, 81, 896–901.

SECTION THREE

PSYCHOSOCIAL ISSUES IN MASTERS SPORT

CHAPTER SEVEN

UNDERSTANDING MASTERS ATHLETES' MOTIVATION FOR SPORT

NIKOLA MEDIC

The objective of this chapter is to discuss motivational processes underlying lifelong involvement in sports. Information pertaining to the topic will be mainly derived from survey data on over 600 Masters Athletes and on archived records from over 40,000 Masters Athletes of different sports; however, additional insights will be provided from other existing motivational research on Masters Athletes. The motivational themes will be discussed with respect to the following:

- The importance of understanding motivational processes underlying lifelong involvement in sport;
- The question of whether relative age effects exist in Masters sports;
- The question of what motivates Masters Athletes.

THE IMPORTANCE OF UNDERSTANDING MOTIVATIONAL PROCESSES UNDERLYING LIFELONG INVOLVEMENT IN SPORT

Over the last three decades, an impressive body of literature has focused on the concept of motivation in sport settings. Various research strategies have been utilized successfully in pursuit of clarifying the complexity of this concept, and over 30 different theories have been proposed to explain and predict what motivates athletes to behave the way they do (Paskevich et al., 2006; Roberts, 2001). This research has been valuable because it provided the basis for understanding the choice, effort, and persistence tendencies, and their relationship to human behaviour. This research has also facilitated the development of strategies for maximizing positive outcomes in sport and physical activity settings in general. Sport motivational literature available to date offers many insights regarding the development of motivational factors until the time athletes reach peak performance in their sports. However, our knowledge is much more limited, but emerging, with regards to the motivational processes of Masters Athletes; that is, individuals who either continue to compete beyond their peak performance, or, at some later time in their life, start or resume training on a daily basis and compete at events available to middle- to older-aged adults (e.g., Master's tournaments, Senior Olympics).

Numerous studies throughout the developed world have shown that, in our aging society, physical activity and sport participation decrease as individuals progress through middle-age and beyond (Dishman, 1994; Grant, 2001). In concert with this, there is convincing evidence that the numerous physiological (Bouchard et al., 2007) and psychological (Biddle et al., 2000) benefits of physical activity and sport involvement outweigh the risks associated with physical activity and sport involvement in older people. Considering that physical inactivity has been highlighted as one of the most important areas for disease-risk-factor reduction in middle- to older-aged adults, efforts have been directed at understanding how to increase and maintain physical activity and sport participation in this population. One approach toward this aim involves trying to understand the motivational processes of a proportionally small, but very unique, sample of the population: Masters Athletes who continue to train for and compete at various sporting disciplines available to older adults. Masters Athletes thus provide us with an exceptional cohort to study motivation for physical activity and sport because they devote a large amount of time to sport and have a lifetime of valuable experience. As such, Masters Athletes are of particular interest because they may have developed and adopted motivational strategies that allow them to maintain sport involvement across the lifespan in spite of age-related performance declines. Moreover, a benefit of utilizing Masters Athletes in research on sport motivation is that they represent the most physically fit and healthy individuals of their cohort, thereby limiting the influence of chronic disease or other physical disability as a barrier to participation. Finally, understanding motivational processes of Masters Athletes is important because extensive evidence suggests that, in order to maintain highly skilled levels of athletic performance, individuals need to engage in adequate amounts of high-quality sport-specific training (e.g., Weir et al. (2002) showed that national-level Masters Athletes trained 6.5 hours per week on average) and must be motivated to overcome setbacks in training and competition over time. Thus, in an attempt to broaden the vision of what constitutes high-performance sport, Masters Athletes should be acknowledged as being representative of the physical elite of an aging population, and select Masters should be recognized as experts who have a lifetime of valuable experience. Spirduso et al. (2005) stated that Masters Athletes:

106

nikola medic

are an important group to study and to emulate . . . because they reveal the limits of human physical potential . . . They are an inspiration . . . because they epitomize optimal physical aging . . . and because they inspire an upward look, provide a standard, and give hope.

(p. 316, see also Horton, Chapter 8)

DO RELATIVE AGE EFFECTS EXIST IN MASTERS SPORTS?

A motivational strategy aimed at establishing a fair playing field in Masters sports involves the use of 'age categories' that generally progress in five-year intervals (e.g., 35–39, 40–44, etc.). The value of this motivational approach is that athletes get to compete against those close to their own age. Age categories are determined by the governing bodies for each Masters sport and are gender specific. However, anecdotal evidence from Masters Athletes suggests that motivational differences exist within each of the five-year age categories. Masters Athletes report that, as they start approaching the upper end of their age category, they feel less motivated to train and compete because of their relative-age disadvantage. For example, in a recent interview in a Masters sports' magazine, Philippa Raschker, a Masters track and field athlete (who, at the time of the interview, held ten world age-group records), was asked the following question: 'Masters Athletes talk a lot about moving into the next age group. How anxious are you about turning sixty?' Her response was:

I turn 59 in February, and due to the injury, this is the perfect time to rest and heal. And yes, I am very anxious to turn 60 and go after the records in that age group. For me, if there were no new goals to achieve, I would change gears and go into another sport. The records are the incentives that have kept me in this program, because they present the challenges I need to pursue after so many years of training.

(Houlihan, 2006, p. 43)

Likewise, a 60-year-old competitive Masters swimmer acknowledged the following:

I had been active in Masters swimming for 20 years and had always managed to establish world records each time I moved into a new age category (every 5 years) and was wondering how I would do at the 60-year-old level . . . I did just that . . . I had gone faster than the old record . . . and I was only a half-second slower than my current world record in the 55–59 year bracket I had set 5 years previously . . . I was very pleased and relieved that I had been able to meet my expectations and the crowd's . . . Each time I "age up", I look forward to setting new standards, and the thrill is always there to race and put myself in the high-pressure zone of tough competition. I don't really know why I do this, but perhaps it is only to gain the recognition of peers and the self-satisfaction of accomplishing a difficult goal.

(Crocker et al., 2004, p. 336–337)

This anecdotal evidence implies that, in addition to the physiological declines related to aging, lack of motivation to train during later stages of the five-year age categories may also contribute to the decline in Masters Athletes' athletic performance.

The motivational effect of age categories in Masters sports, and the influence of Masters Athletes' relative-age advantage/disadvantage within constituent age categories, were systematically examined in a recent study by our research group (Medic et al., 2007). We reasoned that, based on Masters Athletes' birth dates, five-year age categories can be used to identify relatively-older and relatively-younger cohorts of Masters Athletes in the same manner in which relatively-older and relatively-younger individuals have been identified in youth sport settings, which generally use one-year age categories. In this study we tested whether a relative age effect was reflected in participation rates (i.e., participation-related relative age effect) and the performance achieve-ments (i.e., performance-related relative age effect) of Masters Athletes across each constituent and successive year within the five-year age categories. We analyzed archived data on 24,831 participation entries from 1996 to 2005, and 1,160 national records set from 1998 to 2005 at USA Masters championships in track and field and swimming. Based on five-year age categories in which Masters Athletes compete, participation-entry and record-setting ages were each scored separately as frequencies in five separate categories (i.e., Year 1, Year 2, Year 3, Year 4, or Year 5) and were collapsed across all five-year age categories. 'Year 1' included Masters Athletes who were in their first year of any five-year age category when they participated or set a record (i.e., those who were either 35, 40, 45, etc.). Likewise, 'Year 2', 'Year 3', 'Year 4', and 'Year 5' comprised frequencies for participation entries and records set by Masters Athletes who were in the second, third, fourth, and fifth years, respectively, in any five-year age category.

Results of this study (Medic et al., 2007) provided preliminary but strong evidence that relative age effects exist in Masters swimming and track and field. The likelihood of participating in the national championships was significantly higher for Masters Athletes who were in their first or second year, and was lower if they were in their fourth or fifth year of any age category. We also found that the probability of setting a USA Masters record was significantly higher if athletes were in the first year of any five-year age category, and was lower if they were in the third, fourth, or fifth year of an age category.

108

The relative age effects among Masters Athletes can be explained in several ways. One of the explanations relates to the physical capabilities of aging athletes. Research conducted with elderly athletes and healthy sedentary individuals suggests that muscular strength and cardiorespiratory efficiency decline with aging (Donato et al., 2003). Although research has attempted to quantify age-related performance changes for Masters Athletes beyond the age of peak performance (see Chapters 3–6), the actual extent and rate of performance decline seem to depend on many factors including the sport type, event, gender, and experimental design. However, most experts would agree that there is an inevitable age-related decline in performance even when the domain-specific daily training is maintained over many years. If we assume that the extent of the decline is generally one per cent per year after the age of peak performance (Evans et al., 1995; Nessel, 2004), then with each five-year increase in age, we would expect a five per cent increase in performance time. For example, at the 2008 USA Indoor Track and Field Championships, the winning time for men between ages 60–64 in the 800m track event was 2:23.05. Coincidently, this individual was 60 years old, and in the first year of his age category. In five years, theoretically his time would be five per cent slower, or 2:30.20, which would have placed him in seventh place rather than first. This example provides insight into a potential reason why relatively younger Masters Athletes are more likely to achieve more national records than their older peers. It is also possible that Masters Athletes' expectations regarding age-related performance decline influence their actual performance and/or willingness to participate in competitions. In particular, those who do not expect to perform well either do not perform well, or do not even take the opportunity to participate in competition. Another potential explanation of relative age effects in Masters sports relates to Masters Athletes' motivational regulations. Considering that relatively younger Masters Athletes are much more likely to participate in national-level competitions and to set a national record, it is possible that, in comparison to peers within the age category, their intrinsic motivation and perceived competence are higher, and/or their amotivation is lower during that period. As a consequence of this more adaptive motivational profile, relatively younger Masters Athletes may be more likely to participate in national competitions and/or set more national records compared to the relatively older cohort.

How are relative age effects in Masters sports influenced by gender?

The results of our recent study of Masters Athletes from track and field and swimming (Medic et al., in press a) showed that relative age effects in

Masters sport are robust across genders. However, a participation-related relative age effect seems to be stronger for males than females (see Figure 7.1a). A potential reason for why a participation-related relative age effect in Masters sports is stronger in males than females may be that male Masters sports are more competitive; that is, the more competitive the sporting environment, the more likely a relative age effect is to occur. Medic et al. (in press a) suggested that the men's Masters sporting environment is more competitive than women's since the total number of male participants at Masters national-level competitions was about 220 per cent higher than the number of female participants. Thus, having fewer individuals (i.e., a smaller competitive 'pool') competing for the same number of awards would make it less difficult for female Masters Athletes to win awards. Another explanation for why a participation-related relative age effect is stronger in male Masters Athletes may be that male Masters Athletes are more concerned with winning and are more likely to compare their performance with the performance of others (i.e., in normative terms) than female Masters Athletes. Research has shown that male Masters Athletes have a less

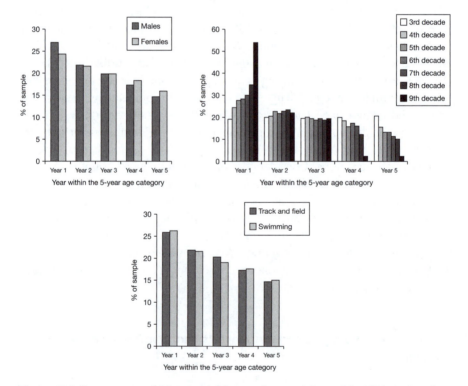

Figure 7.1 Percentage of Masters Athletes who participated in USA national competitions across gender, decades of life, and sport types

self-determined motivational profile and are more ego-oriented than females (Medic et al., 2004; Tantrum & Hodge, 1993). Therefore, because of their stronger emphasis on winning and higher need for social comparison, male Masters Athletes are probably less likely to participate in competitions when chances of setting a record or placing higher in a competition are diminished, as is the case when they are in one of the later years of a five-year age category.

How are relative age effects in Masters sports influenced by age?

The results of our two studies (Medic et al., under review; Medic et al., 2008) showed that participation-related relative age effects in Masters sport begin at the age of 40 years and get progressively stronger with each successive decade of life (see Figure 7.1b). One probable explanation for why this effect becomes more pronounced with age may be because extrinsic rewards are more important and/or more available for older Masters Athletes than for younger ones. Our research has shown that Masters Athletes who are 65 years and older have higher external regulation (i.e., are more likely to be motivated by extrinsic rewards) than Masters Athletes who are between 35 and 64 years (Medic et al., 2004). Also, given that the number of participation entries at Masters national competitions decreases after the fourth decade, and that, at the same time, the number of awards available stays the same, awards are more available to Masters competitors in their 50s and beyond. Thus, because of their stronger emphasis on winning and/or actual and/or perceived chances of winning, older Masters Athletes seem to be more likely to compete at organized national-level events when the chances of setting a record or winning awards are highest — that is, during the time when they are in their first year of an age category.

How are relative age effects in Masters sports influenced by sport type?

Relative age effects exist in Masters-level track and field and swimming. However, *after controlling for participation rates of Masters Athletes*, a performance-related relative age effect seems to be stronger in Masters-level swimming than in track and field (see Figure 7.2c) (Medic et al., 2008; Medic et al., in press a). One potential reason for this may relate to qualifying standards. Specifically, the USA national swimming championships require that qualifying standards be met as a condition of participation (i.e., swimmers can enter up to three events without making qualifying times; however, qualifying times must be met

111

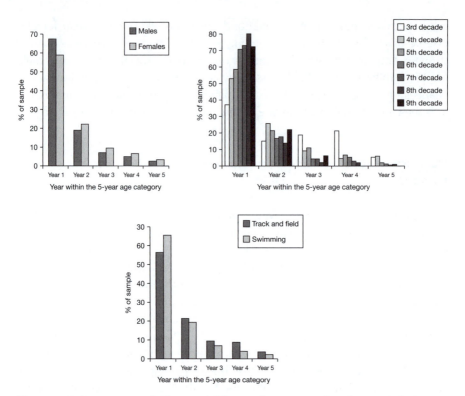

Figure 7.2 Percentage of Masters Athletes who set a national record during USA national championships across gender, decades of life, and sport types

if a swimmer is entering between four and six events), whereas the USA Masters track and field championships do not. This may mean that the Masters swimming environment is more competitive, and as such, one in which a relative age effect is more likely to occur.

In another study (Medic et al., in press b), we replicated prior findings which suggested that a participation-related relative age effect exists in Masters swimming and track and field. However, no evidence for it was found in Masters weightlifting or rowing. In this study, we analyzed 4,820 participation entries from the 2005 World Masters Games. We found that the likelihood of participating in the swimming competition was higher if individuals were in the first year of an age category, and that the likelihood of participating in track and field was higher if individuals were in the first and second year, and lower if they were in the fourth and fifth year of an age category. Participation data for Masters weightlifting and rowing showed that the probability of participating was equally distributed among individuals across all five constituent years of an age category. The results of this study were novel because they showed that participation-related relative age effects may not generalize to all Masters sports; specifically, that such effects do not seem to exist in sports where competitors are arranged by age and weight. For example, in Masters sports such as weightlifting and rowing, in which competitions are arranged by age and weight class, this protocol might serve to 'hide' the perceived age disadvantage for relatively older individuals. As such, the perceived disadvantage that discourages participation might be less evident for Masters competitors arranged by age and weight class than for Masters competitors arranged by chronological age only, as in the case of track and field and swimming.

What are the implications of relative age effect findings in Masters sport?

The findings from studies on relative age effects in Masters sports have several important implications. First, even though relative age differences in Masters sport do not stay with the individuals throughout their athletic careers, as is the case for athletes within youth sports, the current Masters system which uses five-year age registration categories allows Masters Athletes to enjoy the relative age advantage every five years when they enter the next successive five-year age category. Given that successive five-year categories were originally intended to even the playing field and provide an incentive for aging athletes to remain competitive and motivated as they age, our findings suggest that the motivational salience of five-year age categories in the context of older competitive sport seems to be much more questionable for males than females in sports where competitions are organized according to age only, and increasingly questionable as individuals progress from the fourth decade of life onwards. Despite the benevolent intentions of the five-year age category system to motivate athletes and increase participation, there are still deficits in this system as evidenced by the irregular patterns of participation and performance achievements. This irregular or intermittent pattern of participation is problematic since continued engagement in sport has been advanced as a primary reason for high-level sport performance, functionality, and maximal health benefits in middle- and older-aged athletes (Young & Starkes, 2005; Young et al., 2008). Thus, one implication may be that the use of more condensed age categories (e.g., three- or four-year age categories), or the use of age-grading systems in concert with five-year age categories should perhaps be given more serious consideration in terms of how Masters competitions are organised and how Masters Athletes are awarded prizes and recognized for their performances.

113

WHY DO MASTERS ATHLETES PARTICIPATE IN SPORT?

The subject of motivation in sport is concerned with questions such as: Why do individuals participate or drop-out of sport? How can nonparticipants be motivated to start participating in a sport? How can active athletes be motivated to train harder or longer, and/or continue their participation? Deci and Ryan's (2002) self-determination theory is one prominent framework for understanding the social conditions that facilitate or undermine a person's intrinsic and extrinsic motivation. Self-determination theory proposes that different types of motivation exist and that they differ by the degree to which they are self-determined, constituting what is called the self-determination continuum (see Table 7.1).

Early research that examined Masters Athletes' motives for sport were conducted with long-distance runners (e.g., Barrell et al.,1989; Carmack & Martens, 1979; Curtis & McTeer, 1981; Fung et al., 1992) and swimmers (e.g., Dodd & Spinks, 1995; Hastings et al., 1995; McIntyre et al.,1992; Newton & Fry, 1998; Tantrum & Hodge, 1993). These early studies relied heavily on open-ended questionnaire

Table 7.1 Self-determination continuum (adapted from Deci and Ryan, 2002)

Continuum	Motivational regulations	Definition
Self-determination High	Intrinsic motivation to experience stimulation	To experience stimulating sensations (e.g. excitement).
	Intrinsic motivation to accomplish things	For the pleasure and satisfaction experienced when one attempts to accomplish or create something.
	Intrinsic motivation to know	For the pleasure and satisfaction experienced while learning or trying to understand something new.
	Integrated regulation	Because it is coherent with other aspects of one's self.
	Identified regulation	Because it is valued.
	Introjected regulation	Because it is reinforced through internal pressures such as guilt, or emotions related to self-esteem.
	External regulation	Because it is controlled by external sources (e.g. rewards).
Low	Amotivation	Absence of intrinsic and extrinsic motivation, such that one's actions have no control over outcomes.

methodology or single-item questionnaire measures to assess Masters Athletes' motivation for sport. For example, Curtis and McTeer (1981) surveyed 750 Masters long-distance runners and found that they participated in running for the following three reasons: goal achievement (77 per cent of the sample), because they were influenced by others (20 per cent), and because they derived psychological well-being from running (19 per cent). Similarly, Carmack and Martens (1979) conducted a study with 315 runners and found that their main reasons for continuing to engage in running were to maintain fitness, enjoy themselves, participate in competition, control their weight, and feel better. Hastings et al. (1995) asked 700 Masters swimmers about their motives for swimming. Six factors emerged including enjoyment, skill development, fitness, achievement, sociability, and tension release. Tantrum and Hodge (1993) surveyed 40 Masters swimmers from Australia and found that staying in shape, having fun, being fit, and improving skills were reported as the most important motives.

Two investigations employed a case-study approach to the participation motives of Masters Athletes. Langley and Knight (1999) examined the meaning of competitive sport participation for a 68-year-old tennis player and found that successive competitive sport involvement for this individual represented a primary adaptive strategy for coping with the aging process, which enhanced his social relationships, the development of his personal identity, and general propensity for his lifelong physical activity. Similarly, Roper et al. (2003) conducted a study with an 88-year-old internationally ranked male runner. The major motivational themes related to his experience in running included a personal tradition of always being physically active; a perception that his focus on running was mainly about overall fitness and a healthy lifestyle, and that performance outcomes were moderately important; the uniqueness of being a senior-age athlete; and the importance of social support from significant others. Overall, studies have shown that Masters Athletes, regardless of the sport they practice and compete in, have a variety of motives, are very self-determined, are goal oriented, and do not intend to stop participating in sport. Studies show that the most important motives for continuing to train and compete are intrinsic in nature.

By investigating differences in participative motives among various sub-groups of Masters Athletes, studies have shown that motivation for sport differs as a function of gender and age. A summary of studies and their general conclusions are presented next.

How are Masters Athletes' motives for sport influenced by gender?

Ogles et al. (1995) surveyed 610 long-distance runners and found that female runners rated weight concerns, affiliation, psychological coping, life meaning, and self-esteem as more important motives for sport than male runners. Tantrum and Hodge's (1993) study found that male Masters swimmers rated winning as most important and females rated losing weight as most important. Hastings et al.'s (1995) study found that female Masters swimmers rated enjoyment, sociability, and fitness as more important than male swimmers. A similar pattern of findings has emerged in studies conducted by Harris (1981), Leedy (2000), and Masters et al. (1993). Gender differences in motives for sport were also found by Toepell et al. (2004), who surveyed 181 Canadian Masters rowers and found that females rated enjoyment, subjective health, skill improvement, social recognition, and emphasis on participation rather than winning as more important reasons for their sport participation than did males. We (Medic et al., 2005) surveyed 197 Masters track and field athletes and found that females reported higher levels of intrinsic motivation to experience stimulation and lower levels of external regulation than males. In our study (Medic et al., 2006) of 319 Masters Athletes from track and field and swimming, we found that females reported higher levels of self-determined intrinsic (i.e., intrinsic motivation to experience stimulation and intrinsic motivation to accomplish; see Table 7.1) and higher levels of self-determined extrinsic (i.e., integrated and identified regulation; see Table 7.1) motives for sport than males. Overall, the results of the studies that have examined Masters Athletes' motives for sport as a function of gender suggest that female Masters Athletes are likely to give higher importance to reasons related to intrinsic motivation, enjoyment, and health and fitness, and lower importance to extrinsic competition and achievement goals. This suggests that male and female Masters Athletes possess qualitatively different motivational profiles. Specifically, female Masters Athletes endorse more adaptive types of motives for sport, and, in comparison to males, they may be more likely to experience psychological benefits (e.g., in terms of exerted effort, well-being, flow, creativity, and self-esteem) from their participation in Masters sports. Finally, it is interesting to note that gender differences in participative motives among Masters Athletes closely resemble the gender differences in participative motives for sport found among samples of younger athletes. Specifically, youth, adolescent, and young adult male athletes generally report higher levels of extrinsic motivation and lower levels of intrinsic motivation compared to females.

How are Masters Athletes' motives for sport influenced by age?

Harris (1981) showed that older runners were less likely to take up running to help with other sports and were more likely to report higher health benefits from running in comparison to younger groups of runners. Summers et al.

(1983) found that older marathon runners (41 to 61 years) were more concerned with weight control and cardiovascular endurance than younger marathon runners. Dodd and Spinks (1995) found that older Masters Athletes from swimming and track and field (60 years and older) were more extrinsically motivated and had higher need for social approval than any of the younger groups; however, their level of intrinsic motivation was not significantly different from that of younger age groups. In a study of 224 marathon runners, Ogles and Masters (2000) found that older runners (aged 50 years and older) were mainly motivated by health reasons, weight concerns, life meaning, and need for affiliation with other runners, whereas younger runners (aged 20 years and less) were mainly motivated by goal achievement and competition. We (Medic et al., 2004) compared motivational regulations of 64 Masters runners (35 years and above) to 35 university runners (18–31) and found that Masters runners were more externally regulated and less amotivated. In another of our studies (Medic et al., 2006), we surveyed 319 Masters Athletes from track and field and swimming and found that high levels of external regulation is the most important factor that distinguished those 65 years and older from other age groups. Altogether, the results of studies that have examined motives for sport as a function of age seem to suggest that older Masters Athletes place greater importance on extrinsic motives, suggesting that older Masters Athletes have a tendency to be controlled by external sources such as rewards or by constraints that are imposed by others. A potential explanation of why older Masters Athletes are more externally regulated may be related to the available opportunities to obtain external rewards. In particular, an examination of entries in Masters competitions clearly indicates that, after the age of 65, there is a large decrease in the number of competitors. Since the number of awards (e.g., medals) stays the same, the odds of winning increase.

CONCLUSION AND DIRECTIONS FOR FUTURE RESEARCH

The main objective of this chapter was to provide an overview of the literature that examined motivational processes underlying lifelong involvement in sports in hopes of providing insight on how lives can be enriched and enhanced through Masters sports involvement. One implication of the studies reviewed in this chapter is that those interested in enhancing and fostering long-term motivation for sport should aim to develop a sporting environment and training climate that promotes optimally challenging training and competitive activities and events that are intrinsically motivating, that involve emphasis on the development of new skills and techniques, and that involve high levels of internalization of the sport activity. Also, it should be acknowledged that extrinsic reasons for sport participation are not unimportant, but rather that they are at the most moderately important for the majority of Masters Athletes. Extrinsic motives seem to be higher among Masters Athletes who are 65 years or older, male, and/or those in the first year of a five-year age category. However, this does not mean that extrinsic motivation should be promoted, as this type of motivation can be associated with a number of maladaptive consequences. Thus, to be effective, strategies used to motivate Masters Athletes need to be individualized so they complement personal reasons for continuing to train and compete in sport.

While the research conducted on motivational processes of Masters Athletes is valuable and encouraging, additional research is needed to better understand how motivation for sport can be enhanced across the lifespan. One area where research is needed relates to the examination of the changes in motives for sport across the lifespan. Given that Masters Athletes can have a large age range, and the fact that the cross-sectional studies suggest that Masters motives differ across age, it remains unclear whether motives for sport change over time. Longitudinal studies are needed to shed light on this issue. In addition, factors that contribute to motivational differences between younger and older Masters Athletes (or between male and female Masters Athletes) have not been investigated and thus remain unknown at this time.

Another area where research is needed relates to the psychological, social, and physiological mechanisms that may explain relative age effects in Masters sports. For example, studies could examine whether Masters Athletes believe that they have specific advantages/disadvantages during the five constituent years of a five-year age category. Future studies may also consider examining whether Masters Athletes' perceptions of competence and motivational regulations differ and/or change across the five constituent years within a five-year age category. Cross-sectional design, especially in cases when the participants are recruited during competitions, is likely to produce a sampling bias since Masters Athletes who are in the last two years of any five-year age category at the time of the assessment would not have an equal chance of being assessed because they are less likely to participate in competitions in the first place. Thus, it would be important that a longitudinal design be employed during which Masters Athletes would be followed and reassessed during each of the five constituent years of a five-year age category. Finally, it would be interesting to determine whether an early-life relative age effect generalizes or transfers across successive age categories such that it manifests in mid-life competitive age categories. Future research could employ retrospective questionnaires to track the participation patterns of athletes leaving adolescence into early adulthood to provide answers in this respect.

REFERENCES

Barrell, G., Chamberlain, A., Evans, J., Holt, T., & Mackean, J. (1989). Ideology and commitment in family life: A case study of runners. *Leisure Studies*, 8, 249–262.

Biddle, S.J.H., Fox, K.R., & Boutcher, S.H. (2000). *Physical activity and psychological well-being.* London: Routledge.

Bouchard, C., Blair, S.N, & Haskell, W.L. (2007). *Physical activity and health.* Champaign, IL: Human Kinetics.

Carmack, M.A., & Martens, R. (1979). Measuring commitment to running: A survey of runners' attitudes and mental states. *Journal of Sport Psychology*, 1, 25–42.

Crocker, P.R.E., Kowalski, K., Hoar, S., & McDonough, M.H. (2004). Emotions in sport across adulthood. In M. Weiss (Ed.), *Developmental sport and exercise psychology: A lifespan perspective* (pp. 333–356). Morgan Town, WV: Fitness Information Technology.

Curtis, J., & McTeer, W. (1981). The motivation for running. *Canadian Runner*, 1, 18–19.

Deci, E.L., & Ryan, R.M. (Eds.), (2002). *Handbook of self-determination research.* Rochester, NY: University of Rochester Press.

Dishman, R.K. (1994). *Advances in exercise adherence.* Champaign, IL: Human Kinetics.

Dodd, J.R., & Spinks, W.L. (1995). Motivations to engage in masters sport. *ANZALS Leisure Research Series*, 2, 61–75.

Donato, A.J., Tench, K., Glueck, D.H., Seals, D.R., Eskurza, I., & Tanaka, H. (2003). Declines in physiological functional capacity with age: A longitudinal study in peak swimming performance. *Journal of Applied Physiology*, 94, 764–769.

Evans, S.L., Davy, K.P., Stevenson, E.T., & Seals, D.R. (1995). Physiological determinants of 10 km performance in highly trained female runners of different ages. *Journal of Applied Physiology*, 78, 1931–1941.

Fung, L., Ha, A., Louie, L., & Poon, F. (1992). Sport participation motives among veteran track and field athletes. *ICHPER*, 24–28.

Grant, B.C. (2001). You're never too old: Beliefs about physical activity and playing sport in later life. *Society and Leisure*, 25, 285–302.

Harris, M. (1981). Runners' perception of the benefits of running. *Perceptual and Motor Skills*, 52, 153–154.

Hastings, D.W., Kurth, S.B., Schloder, M., & Cyr, D. (1995). Reasons for participating in a serious leisure career: Comparison of Canadian and U.S. masters swimmers. *International Review for the Sociology of Sport*, 30, 101–117.

Houlihan, M. (2006, February/March). A masterful athlete. *GeezerJock*, 3, 42–43.

Langley, D.J., & Knight, S.M. (1999). Continuity in sport participation as an adaptive strategy in the aging process: A lifespan narrative. *Journal of Aging and Physical Activity*, 7, 32–54.

Leedy, M.G. (2000). Commitment to distance running: Coping mechanism or addiction? *Journal of Sport Behavior*, 23, 255–270.

Masters, K.S., Ogles, B.M., & Jolton, J.A. (1993). The development of an instrument to measure motivation for marathon running: The motivations of marathoners scale (MOMS). *Research Quarterly for Exercise and Sport*, 64, 134–143.

McIntyre, N., Coleman, D., Boag, A., & Cuskelly, G. (1992, Summer). Understanding masters sports participation: Involvement, motives and benefits. *The ACHPER National Journal*, 4–8.

Medic, N., Starkes, J.L., & Young, B.W. (2007). Examining relative age effects on performance achievement and participation rates of masters athletes. *Journal of Sports Sciences*, 25(12), 1377–1384.

Medic, N., Starkes, J.L., Young, B.W., & Weir, P.L. (2006). Motivation for sport and goal orientations in Masters athletes: Do masters swimmers differ from masters runners? *Journal of Sport and Exercise Psychology*, 28, s132.

Medic, N., Starkes, J.L., Young, B.W., Weir, P.L., & Giajnorio, A. (2004). Masters athletes' motivation: Evidence of age and gender differences. *Canadian Society for Psychomotor Learning and Sport Psychology Abstracts*, Saskatoon, Saskatchewan.

Medic, N., Starkes, J.L., Young, B.W., Weir, P.L., & Giajnorio, A. (2005). Multifaceted analyses of masters athletes' motives to continue training and competing. *ISSP 11th World Congress of Sports Psychology*, Sydney, Australia, 1, 1–3.

Medic, N., Young, B.W., & Saarloos, D. (2008, November). Promoting physical activity to middle and older aged adults: Examination of a recurring relative age effect across the lifespan in Australian sport. *Australian Association of Gerontology: 41st National Conference*. Fremantle, Western Australia.

Medic, N., Starkes, J.L., Weir, P.L., Young, B.W., & Grove, J.R. (in press a). Gender, age, and sport differences in the relative age effects among USA Masters swimming and track and field athletes. *Journal of Sport Sciences*.

Medic, N., Starkes, J.L., Weir, P.L., Young, B.W., & Grove, J.R. (in press b). Relative age effect in masters sports: Replication and extension. *Research Quarterly for Exercise and Sport*.

Nessel, E.H. (2004). The physiology of aging as it relates to sports. *Journal of the American Medical Athletic Association*, 17, 12—17.

Newton, M., & Fry, M.D. (1998). Senior Olympians' achievement goals and motivational responses, *Journal of Aging and Physical Activity*, 6, 256–270.

Ogles, B.M., & Masters, K.S. (2000). Older vs. younger adult male marathon runners: Participative motives and training habits. *Journal of Sport Behavior*, 23, 130–142.

Ogles, B.M., Masters, K.S., & Richardson, S.S. (1995). Obligatory running and gender: An analysis of participative motives and training habits. *International Journal of Sport Psychology*, 26, 233–248.

Paskevich, D.M., Dorsch, K.D., McDonough, M.H., & Crocker, P.R.E. (2006). Motivation in sport. In P.R.E. Crocker (Ed.), *Introduction to sport psychology: A Canadian perspective* (pp. 72–101). Toronto, Ontario: Pearson Education Canada.

Roberts, G.C. (2001). *Advances in motivation in sport and exercise*. Champaign, IL: Human Kinetics.

Roper, E.A., Molnar, D.J., & Wrisberg, C.A. (2003). No old fool: 88 years old and still running. *Journal of Physical Activity and Aging*, 11, 370–387.

Spirduso, W.W., Francis, K.L., & MacRae, P.G. (2005). *Physical dimensions of aging* (2nd ed). Champaign, Illinois: Human Kinetics.

Summers, J.J., Machin, V.J., & Sargent, G.I. (1983). Psychological factors related to marathon running. *Journal of Sport Psychology, 5,* 314–331.

Tantrum, M., & Hodge, K. (1993). Motives for participating in masters swimming. *New Zealand Journal of Health Physical Education and Recreation, 26,* 3–7.

Toepell, A.R., Guilmette, A.M., & Brooks, S. (2004). Women in Masters rowing: Exploring healthy aging women's health and urban life. *International and Interdisciplinary Journal, 3,* 74–95.

Young, B.W., & Starkes, J.L. (2005). Career-span analyses of track performance: Longitudinal data present a more optimistic view of age-related performance decline. *Experimental Aging Research, 31,* 1–21.

Young, B.W., Weir, P.L., Starkes, J.L., & Medic, N. (2008). Does lifelong training temper age-related decline in sport performance? Interpreting differences between cross-sectional and longitudinal data. *Experimental Aging Research,* 34(1), 27–48.

Weir, P.L., Kerr, T., Hodges, N.J., McKay, S.M., & Starkes, J.L. (2002). Master swimmers: How are they different from younger elite swimmers? An examination of practice and performance patterns. *Journal of Aging and Physical Activity, 10,* 41–63.

CHAPTER EIGHT

MASTERS ATHLETES AS ROLE MODELS?

Combating Stereotypes of Aging

SEAN HORTON

Recently the Heart and Stroke Foundation issued a cautionary warning to the 'baby-boom' generation, of which one-half are sedentary and almost one-third are obese. The foundation warned of severe consequences with respect to the health and well-being of this cohort as their oldest members began turning 60 years of age. The problem appears to be one of action rather than one of knowledge. Ninety-eight per cent of people over the age of 50 are aware that physical activity is important to maintaining their health (Ory et al., 2003), yet only a minority of senior women and men get sufficient physical activity to maintain optimal health benefits (Statistics Canada, 2005).

One potential barrier to seniors' participation in physical activity is prevailing cultural attitudes and stereotypes, which in North America tend to be predominantly negative towards seniors (Levy & Langer, 1994). Researchers (e.g., Levy & Myers, 2004; O'Brien Cousins, 2000) have suggested a link between aging stereotypes and the lack of physical activity of older adults, which ultimately affects long-term health. Seniors often buy into the negative stereotypes, which influence their decisions to get involved, or stay involved, in various activities.

How do we break this negative cycle? Masters Athletes are generally considered to be some of society's most successful agers due to the fact that they maintain very high levels of performance into the latter stages of life. Consequently, Masters Athletes are studied by researchers and frequently profiled in the popular press for their athletic achievements. Indeed, their ability to 'bust' popular stereotypes of aging is part of what makes their stories so compelling. In many respects, Masters Athletes would appear to be ideal role models both for society as a whole and for other seniors. While the research into role models specific to seniors is in its early stages, the evidence suggests the complexity inherent in these issues. Seniors react to Masters Athletes in diverse and multifaceted ways, which raises interesting questions with respect to motivating people to engage in exercise.

SOCIETAL STEREOTYPES OF SENIORS

Historically, we have not been a culture that honors or reveres our elderly. Images and stereotypes of aging throughout history have been predominantly negative. In ancient Greece, the literature and writings tended to portray old age as disgusting, ugly, and tragic (Gilleard, 2007). William Shakespeare, writing *As You Like It* in 1607, described the seventh and final stage of life as a reversion to childhood, and eventually to complete oblivion, without teeth, without eyes, without taste — indeed, 'sans everything'.

One might argue that portrayals of seniors are not much different in the twenty-first century. Levy and Banaji (2002) suggested that stereotypes of the elderly continue to be predominantly negative, which has consequences for how seniors are treated. Ory et al. (2003) noted that 91 per cent of seniors in Canada and the United States have experienced episodes of ageism, with more than 50 per cent experiencing multiple incidents. Many of these were of a subtle nature — receiving a birthday card that made fun of older people, for example. This reflects the restrained character of ageism in North America; the kind of explicit hatred and severe discrimination that other stigmatized groups have had to endure are rare. Instead, the discrimination that seniors face comes in the form of being ignored, or having health care providers attribute ailments to their age (Ory et al., 2003).

Ory et al. (2003) provided some evidence of a subtle shift, suggesting that increased access to health care and improving economic circumstances in the past 30 years have helped to improve the images and stereotypes of seniors. The recent plethora of pharmaceutical advertisements displaying seniors as active and attractive supports such an assertion. Overall, however, the authors stated that popular stereotypes of aging remain predominantly negative.

Disturbingly, seniors themselves tend to buy into the negative stereotypes. Montepare and Zebrowitz (2002) noted that seniors see members of their own group as less goal oriented, less likeable, unhappier, and more dependent than younger adults. Perhaps this is not surprising. By the time individuals reach their senior years, they have spent the majority of their lives expressing and internalizing negative stereotypes of the elderly. It may be inevitable that seniors' views of their own social group will be as negative as those the rest of society holds. As Levy and Banaji (2002) remarked in their discussion of the elderly and their tendency towards negative self-characterizations: with friends like oneself, who needs enemies?

Seniors' tendencies to negatively self-stereotype were further illustrated by research into implicit attitudes (Nosek et al., 2002). The Implicit Association Test (IAT) compares seniors' implicit (or unconscious) attitudes with their explicit (or consciously expressed) views. The IAT measures automatic attitudes and stereotypes of a number of different social groups, including the elderly. The test itself pairs a social group (i.e., old–young) with an evaluative dimension (i.e., good–bad). The response latency, or the speed at which pairings are made, is a measure of the strength of the implicit attitude (the IAT can be found at www.yale.edu/implicit). Data were collected over a three-year period from 68,000 individuals, ranging in age from eight to 71.

As expected, both explicit and implicit attitudes towards the elderly held by young people were negative. As the age of the respondents increased, explicit (or conscious) attitudes towards the elderly improved, albeit modestly. Implicit (unconscious) attitudes, however, remained consistently and overwhelmingly negative irrespective of age. In fact, implicit attitudes towards the old were as negative among the elderly as among the young. Nosek et al. (2002) noted that members of other groups tested (i.e., on race or gender) generally showed more positive implicit attitudes towards their own group compared to non-group members. The elderly seem to be an exception to this trend.

IMPACT OF STEREOTYPES ON PERFORMANCE, HEALTH, AND WELL-BEING

There appear to be important consequences of negative stereotyping. The evidence suggests that negative societal attitudes, along with the destructive self-stereotypes that seniors possess, affect a myriad of areas, including performance on physical and cognitive tasks and decisions to engage in physical activity, as well as the overall health and well-being of our senior population.

While the IAT examines implicit attitudes, or subconscious bias, Levy et al. have conducted a number of experiments that demonstrate how subtle manipulations at a subconscious level can affect a variety of performance measures. For example, Levy (1996) exposed seniors to either negative or positive words of aging while they were ostensibly playing a computer game. These stereotypical words of aging were flashed at speeds fast enough to bypass conscious awareness; thus, participants were unaware that they were being 'primed'. Seniors who were exposed to negative words (e.g., senile, Alzheimer's) performed worse on a

124

subsequent memory task than those who received positive words (e.g., wise, sage).

Hausdorff et al. (1999) utilized a similar priming methodology to examine the effect on walking speed of seniors. They found that seniors, after being primed for 30 minutes with positive aging words, improved both their walking speed and swing time (the time with one foot in the air during walking). Those who were negatively primed showed no change in these two measures. Studies examining handwriting (Levy, 2000) and cardiovascular response (Levy et al., 2000) have shown these areas to be similarly sensitive to subconscious priming.

A second body of research has examined the effect of explicit, (i.e., conscious) stereotyping on seniors (e.g., Chasteen et al., 2005; Desrichard & Kopetz, 2005; Hess et al., 2003). The majority of these studies have examined how stereotypes influence memory performance. For example, Hess et al. (2003) had one group of participants read a fictional news article that emphasized the negative aspects of aging, while a second group read a fictionalized article that emphasized the positive aspects of aging. All participants were subsequently given a memory task. The authors found that, for the participants who were highly invested in their memory skills, the negative stereotype depressed their scores. Hess et al. (2003) hypothesized that the negative condition evokes a 'stereotype threat' which impedes one's ability to perform, particularly on tasks that are related to self-identity or self-esteem.

While much of the research on stereotypes has examined the short-term implications for performance, the more insidious effects may occur over a longer period of time. Steele (1997) considered 'disidentification' to be the most serious long-term effect of chronic exposure to negative stereotypes. Steele's research on black students led him to conclude that continued exposure to negative stereotypes regarding their intellectual abilities led many to disidentify with academic achievement. Disindentification involves a process of removing a particular domain from one's self-identity. Steele hypothesized that black students often remove intellectual achievement from their self-identity in order to protect their self-esteem. In the same manner, women may disidentify with science- or math-based careers based upon negative stereotypes of women in those particular fields. Seniors, when faced with consistent negative stereotyping regarding their physical and cognitive functioning, may start to avoid activities that challenge these skills.

While disidentification can serve a protective function, it is also likely to decrease motivation, thereby contributing to long-term performance decline in these areas. Seniors who disidentify with, and ultimately remove, cognitive and physical skills as a basis for self-evaluation are likely to experience a downward spiral in these areas (Whitbourne & Sneed, 2002). Maharam et al. (1999) maintain that a significant portion of the physical decline that we have come to see as 'normal' in elderly populations is largely due to sedentary lifestyles rather than an inevitable biological process. Research in domains as varied as typing (Salthouse, 1984), golf (Baker et al., 2005), chess (Charness, 1981), and piano (Krampe & Ericsson, 1996) all suggest that performance can be retained to a remarkable degree with continued involvement in the domain. The danger inherent in negative aging stereotypes is that they influence beliefs about aging, along with beliefs about what is possible and appropriate in later life. Ultimately, these beliefs may affect long-term health outcomes by influencing health behaviors (Sarkisian et al., 2005).

O'Brien Cousins (2000; 2003) examined the manner in which aging beliefs affected engagement in physical activity. In a study that assessed older women's responses to six different exercises (e.g., brisk walking, aquacise, riding a bike or cycling), the author found that women often had reasons why they should not take part in such activities (O'Brien Cousins, 2000). While most respondents recognized the benefits of engaging in exercise, the conceptions of the risks involved often outweighed the perceived benefits. The author found that beliefs about negative outcomes arising from these exercises were 'pervasive, strong, and even sensational in description' (O'Brien Cousins, 2000, p. 291).

It appears that one's beliefs can affect a multitude of areas related to overall health. Levy and Myers (2004) found that seniors with more positive expectations of the aging process were more likely to engage in regular exercise, and also more likely to engage in other behaviors related to good health, including eating a balanced diet and making regular visits to their doctor.

In general, older adults are more likely to attribute their ailments to their advanced age, and seek treatment less assertively than younger adults (Ory et al., 2003). This is compounded by actions of medical professionals. The beliefs held by doctors, nurses, and others in the health care industry can have a significant impact on the treatment and overall health of the elderly. Doctors tend to provide less aggressive treatments to older patients, irrespective of how patients would fare with those treatments (Giugliano et al., 1998). In addition, ageist attitudes often affect the quality of doctor–patient communication, which can result in suboptimal treatments and follow-up care (Adelman et al., 2000). Bowling (1999), in an analysis of ageism affecting patients with cardiovascular disease, concluded that 'ageism in clinical medicine and health policy reflects the ageism evident in wider society' (p. 1353).

In a sense, ageist attitudes and negative stereotypes of aging that exist in society result in a 'double-whammy' to seniors. First, they influence the manner in which seniors are treated by society as a whole, affecting areas as diverse as the way seniors are depicted in children's literature (e.g., Dellmann-Jenkins & Yang, 1997), to the quality of medical care that they receive. Second, and perhaps more importantly, cultural stereotypes affect how seniors see themselves. The ramifications here are enormous, for research shows that stereotypes influence decisions to engage in cognitive and physical activity, as well as decisions around health care, all of which ultimately affect the length and quality of life (Levy et al., 2002). Considering current demographic trends and the sheer number of people about to embark on their senior years, finding ways to effectively minimize and counteract the most negative aspects of aging stereotypes remains a pressing social concern.

COMBATING THE STEREOTYPES

Levy and Banaji (2002) noted that exemplary individuals have the potential to change attitudes towards a social group. It is feasible that seniors who maintain a very high level of health and functioning into late ages might affect both societal attitudes in general, and the beliefs of seniors themselves. Masters Athletes and others who accomplish remarkable athletic feats relatively late in life often appear in the media, and are held up as examples of what is possible in one's senior years. There is little in the way of systematic research, however, as to how such individuals influence aging stereotypes held by society as a whole. An examination of the manner in which Masters Athletes may affect societal stereotypes and perceptions of aging is an intriguing topic of future research.

There has been some investigation as to how highly active seniors will affect others specific to their own age cohort. Ory et al. (2003) noted that, ironically, elite senior athletes may discourage other seniors from engaging in physical activity. The authors stated that elite athletes are likely to intimidate rather than inspire seniors to be more active in their own lives. Ory et al. indicated that portraying 'real people' taking part in 'realistic activities' were the images that were most effective in motivating seniors to be more active.

Recent work (Horton et al., 2008) has added some complexity to these assertions. In interviews we conducted with 20 seniors (aged 62–74), participants were asked their opinions of an elite marathon runner, Ed Whitlock. Whitlock has the distinction of being the oldest participant to run a marathon in less than three hours, which he most recently accomplished at age 72. Participants were shown a picture of Whitlock, and told of his various accomplishments and exercise regime. The participants responded in a variety of ways to Whitlock, but responses generally fell into one of three distinct categories.

Group one responded in a manner consistent with that predicted by previous research; they found that this elite runner was indeed intimidating, and not likely to inspire them to engage in more physical activity. Typical responses from participants in this group include statements such as:

> I've never been that athletic or that competitive, you know, so in a sense I don't really identify with him.

> I think they're really, really pushing it. You know, I can see reasonable exercise, but I can't see doing marathons when you're that age.

A second group of participants saw this elite athlete as fairly extreme yet potentially motivating and inspiring for a certain segment of seniors.

> For someone like me, yes, but I think for the average senior, maybe not. The average senior doesn't aspire to that level of fitness. I would love to be that, and be able to do that.

> I would think so, but he may be too far out. Other people might say, 'I could never do that'. Now, they might be able to if they trained properly.

> To a lot of seniors, they'd say, 'Well, isn't that nice? But I could never do that', and they wouldn't even try, and they'd be a little bit overawed, I think.

Finally, a third group of participants was unequivocal in their admiration of this elite athlete, noting that he would be a viable and appropriate role model for seniors.

> I admire him. I wish I could do it. I think it's marvelous, absolutely marvelous. I admire the man.

> (Would he be an appropriate role model for seniors?)

> Yes. Yeah, 'Get off your butts and get busy.'

Another participant commented, 'I think that's just amazing. Well, I could picture myself, if I didn't develop arthritis, still running at that age.'

While these findings partially support those of earlier research, they also add some complexity, as it appears that elite athletes like Ed Whitlock can be

128

appropriate models for a certain segment of seniors. Overall, there was a tendency for seniors who were more physically active to view Whitlock as a potential role model. There were, however, exceptions to that general trend — seniors who were relatively sedentary and saw him as a viable role model, and active seniors who did not. In fact, one senior, who reported walking up to two hours per day and working out with weights on a daily basis, indicated that he found the picture of the marathoner distinctly unappealing: 'I look at him as an aberration. I mean . . . to look at the picture to me is almost stressful to look at. No, that doesn't do anything for me. It turns me right off'.

At the outset of the interviews, prior to seeing the picture of Ed Whitlock, participants were asked about their own personal role models for the aging process, and what aging successfully meant to them. Most of the participants had people in their lives who provided inspiration. Generally these people were friends or family, and occasionally individuals who had been profiled in the media. Normally their role models were 10 to 20 years older than the participants themselves and, in spite of this advanced age, in better physical condition.

I've got an uncle — he's in his mid-80s. And he is more active now than I have ever been or ever will be, I think. I don't try to emulate him, but I do try to — I do admire him and I do try to stay relatively active compared to him. This guy kayaks, he builds these big sea kayaks, and he has kayaked around Newfoundland, all around the whole Island . . . not in one stop. But he'd kayak and camp. He's just amazing.

There was a guy in the congregation who was 91. I asked for volunteers to come and help me move from one apartment to the other. And he was the only one who showed up, and I thought, 'Oh boy'. And he was fully dressed, 91, shirt, tie, vest, coat, running shoes. And we went for the whole day; even after I was tired, he was still going. And I guess he influences me, because I think of him from time to time. I said, 'How do you stay in this kind of shape? You're in better shape than I am'.

I have a friend that I golf with. He's 79 years old, he walks the course, he exercises. He still goes to the Y. I consider him to be a role model . . . he's in extremely good physical shape. Not only for a man his age, but for a man 20 years younger than him, and I think of him as a physical role model, because he can walk the course and I can't, and I'm nine years younger than him.

I have a cousin, he just turned 80 and he's still playing senior hockey. He's the type of fellow who's always kept himself in quite good shape —

I don't think he ever smoked, never was overweight, always was active, a moderate, social drinker. He certainly doesn't look 80. So I always admired him and still do.

Invariably, the individuals that participants mentioned were active, vigorous, and enjoying a high quality of life into their latter years. They were generally family or friends, which supported the findings of Lockwood et al. (2005), who noted that seniors are most likely to pick family, friends or acquaintances as health-related role models. Occasionally someone in the media caught their attention, one example being a woman skydiver who was profiled in a local newspaper: 'There was this woman who jumped out of a plane on her 90th birthday . . . amazing'.

It is noteworthy that many of the individuals that participants mentioned as personal role models fit the profile of a Masters Athlete. While participants had a mixed reaction to Ed Whitlock (whose running accomplishments make him an outlier, even by Masters Athlete standards) the majority of role models that participants cited were very fit, and engaged in either sport or physical activity of some kind.

Bandura (1977) emphasized the potential of such role models in fostering behavior change. Bandura's social cognitive theory outlined the importance of self-efficacy — a situation-specific form of self-confidence. For example, a golfer may have a high level of confidence in her abilities overall; however, in tournament conditions, in poor weather, playing on a course she has never played before, her level of self-efficacy for playing well that particular day may suffer. Social cognitive theory describes how modeling, or 'vicarious experiences', can be effective in instilling a belief in one's ability to reach a goal or objective, particularly if the model is perceived as being similar to oneself. While it is usually most effective to have a model who is known to the individual (i.e., a friend or acquaintance), often people who are not known personally can serve as role models. For example, a celebrity who starts an exercise program can serve as a motivator for an individual to start their own exercise regime (Lox et al., 2003). A second tenet of Bandura's (1977) social cognitive theory relevant to modeling is 'social persuasion'. Often a person who is considered knowledgeable (i.e., a doctor, or someone considered to be an expert) is particularly effective in increasing self-efficacy in others.

These two tenets of social cognitive theory — social persuasion and vicarious experiences — may be relevant with respect to the potential impact of Masters Athletes on other seniors. Masters Athletes may have specific knowledge with respect to exercise, training, and diet that could be shared with others, and, as 'experts' in this area, could help to encourage and motivate other seniors to engage in more healthful behaviors. Profiling Masters Athletes in newsletters targeted at seniors, or developing workshops where they can share their stories and experience, are potential methods of utilizing that expertise. Finding ways for Masters Athletes to lend their knowledge, and at the same time neutralizing the intimidation factor, would enable these athletes to make an important contribution to the health of those in their community.

Finding appropriate role models is a delicate balancing act. Lockwood and Kunda's (1997) investigation into the effect of 'superstars' on motivation suggested that some superstars can discourage rather than motivate. The authors measured how people responded to elite performers in a domain that held particular interest. They concluded that the ideal role model was someone who was slightly older and had achieved outstanding but not impossible success in an area in which an individual hoped to excel. Two conditions appeared to act as de-motivators for the study participants: when the superstar model was of the same age, and thus their success was perceived to be unreachable due to the fact that it was already 'too late'; or when the superstar's success was so extraordinary it was similarly perceived to be out of reach.

It is important to note that this research was conducted with younger adults and requires replication in an older population; but it does suggest that elite senior athletes may turn older people off exercise, particularly if the athlete is of the same or similar age, and their success is deemed too extreme to emulate. Elite Masters Athletes may be most suitable and provide the greatest motivation to younger generations, particularly to those who view that kind of athletic involvement as feasible and attainable. While elite Masters Athletes may be effective in altering societal stereotypes and inspiring those in a younger cohort, they could have a very different effect on those in their own peer group.

Subsequent work by Lockwood et al. (2005) found that, as individuals age, a gradual shift occurs from a focus on health *promotion* to a more balanced *prevention-promotion* approach. As a result, seniors tended to be more 'loss-focused' and less 'gain-focused' than younger adults. While younger adults looked primarily to positive health models for inspiration, older adults used both upward and downward social comparisons. Older adults may continue to find positive health role models motivational and inspiring, but there is an increasing tendency to compare downwards as one ages. Downward comparisons (e.g., an individual suffering from emphysema due to a lifetime of smoking) can motivate people to take action in an attempt to avoid the same negative consequences. This type of regulatory focus (i.e., one's prevention versus promotion orientation) would appear to have major implications for behavior change. One important area of future research is determining how people's regulatory focus may influence the effectiveness of role models.

Complicating these notions is the type of behavior that people are attempting to change. Lockwood et al. (2005) hypothesized that downward comparisons may be more effective for changing a negative behavior. For someone who wishes to stop smoking, a 'model' who displays the long-term effects of smoking may be the most effective way of fostering a change in behavior. For someone who wishes to start an exercise program, however, an upward social comparison — someone who has initiated a program and is deriving health benefits as a result — may be more effective. Overall, this leaves intriguing and complicated questions for future research.

SUMMARY AND FUTURE RESEARCH

While research in this area is in its preliminary stages, there are some tentative conclusions we can draw from existing studies. Even as Masters Athletes may be inspirational and viable role models for younger generations, potentially altering societal conceptions of growing old, their effect on their peer group is more equivocal. Some peers will see elite athletes as realistic role models and find inspiration in their example. Others, however, will see these individuals as too extreme and their accomplishments unattainable, potentially decreasing their motivation to engage in physical activity.

There appear to be distinct links between one's expectations of the aging process and the likelihood of engaging in exercise. While Lockwood et al. (2005) noted that seniors generally expect a decline in their future selves over the next 10–15 years of their lives, work by Levy and Myers (2004) suggests there is likely to be considerable individual variability in this respect. Levy and Myers found that seniors with higher expectations of their own aging took better care of their health, which included eating a healthier diet and getting more exercise. The relationship between expectations of aging and exercise was further reinforced by Sarkisian et al. (2005), who found that just eight per cent of seniors who had engaged in less than one hour of physical activity over a seven-day time frame had high expectations of the aging process. For those seniors who had exercised for more than three hours in the past week, however, 30 per cent had high expectations of their own aging. The authors suggested that an intervention aimed at raising seniors' expectations of aging could lead to healthier lifestyles. Alternatively, it is feasible that exercise itself results in higher expectations of aging. Future research will need to look at the relationship between physical activity levels, expectations of the aging process, and how that may influence the effectiveness of various role models.

132

Research on the role models for older adults has focused primarily on seniors with a relatively high degree of functioning and living independently in their community (Lockwood et al., 2005). It is possible that viable role models will vary dramatically depending on the specific health concerns with which a person is living. In addition, research has focused on seniors who are relatively young — from 60 to 75 years of age. There is a distinct lack of investigation into the role models for those who are considered to be 'middle-old' (from 75 to 90 years of age) or 'old-old' (90+). Estimates suggest that the numbers in these two latter age groups will increase dramatically in coming years. Even very conservative estimates predict a tenfold increase in the number of centenarians by 2050 (Payne & Issacs, 2008). The implications of such demographic trends are profound for areas such as public pensions and health care; finding ways of maintaining physical functioning and a high quality of life for people in these age categories takes on added importance and urgency. The evidence suggests that exercise programs are beneficial irrespective of age; individuals in their 90s have shown significant strength improvements as a result of training (Fiatarone et al., 1990). Questions remain as to how role models may change as one moves from 'young-old' to 'old-old', and whether there is a further shift in the prevention-promotion orientation. Understanding what motivates the 'oldest-old' to engage in exercise, along with the barriers they may face, is an important area of inquiry.

CONCLUSION

Cultural stereotypes of aging affect seniors in a number of predominantly negative ways. Research has shown that negative aging stereotypes can affect cognitive and physical performance of seniors, recovery from disease, and even longevity. Levy and Banaji (2002) noted the importance of challenging these negative stereotypes, and that exemplars or role models can potentially play an important part in changing societal perceptions of aging. Interviews conducted with seniors showed how their role models represent what it means to age successfully. Invariably these were individuals older than themselves, active, vigorous, and illustrative of the high quality of life that is possible into a very late age. Masters Athletes tend to fit this description. While this is a relatively new and under-explored area of research, it is conceivable that these athletes can play an important part in (a) changing societal expectations towards physical activity and aging, and (b) positively affecting seniors' attitudes towards exercise. While seniors react to exceptional individuals in their age group in multifaceted ways, and while there are possible drawbacks to using 'elite' athletes as role models, the potential exists for Masters Athletes to inspire and increase motivation to

133

partake in physical activity. Given that the vast majority of seniors in North America fall short of meeting the minimum daily physical activity requirements for maintaining health, a focus on Masters Athletes may translate into important societal benefits.

REFERENCES

Adelman, R.D., Greene, M.G., & Ory, M.G. (2000). Communication between older patients and their physicians. *Clinical Geriatric Medicine*, 16, 1–24.

Baker, J., Horton, S., Pearce, W., & Deakin, J. (2005). A longitudinal examination of performance decline in champion golfers. *High Ability Studies*. 16, 179–185.

Bandura, A. (1977). *Social learning theory*. Englewood Cliffs: Prentice Hall.

Bowling, A. (1999). Ageism in cardiology. *British Medical Journal*, 319, 1353–1355.

Charness, N. (1981). Search in chess: Age and skill differences. *Journal of Experimental Psychology: Human Perception and Performance*, 7, 467–476.

Chasteen, A., Bhattacharyya, S., Horhota, M., Tam, R., & Hasher, L. (2005). How feelings of stereotype threat influence older adults' memory performance. *Experimental Aging Research*, 31, 235–260.

Dellmann-Jenkins, M., & Yang, L. (1997). The portrayal of older people in award-winning literature for children. *Journal of Research in Childhood Education*, 12, 96–100.

Desrichard, O., & Kopetz, C. (2005). A threat in the elder: The impact of task instructions, self-efficacy and performance expectations on memory performance in the elderly. *European Journal of Social Psychology*, 35, 537–552.

Fiatarone, M.A., Marks, E.C., Ryan, N.D., Meredith, C.N., Lipsitz, L.A., & Evans, W.J. (1990). High intensity strength training in nonagenarians: Effect on skeletal muscle. *Journal of the American Medical Association*, 263, 3029–3034.

Gilleard, C. (2007). Old age in ancient Greece: Narratives of desire, narratives of disgust. *Journal of Aging Studies*, 21, 81–92.

Giugliano, R.P., Camargo, C., Lloyd-Jones, D., Zagrodsky, J., Alexis, J., Eagle, K., Fuster, V., & O'Donnell, C.J. (1998). Elderly patients receive less aggressive medical and invasive management of unstable angina: potential impact of clinical guidelines. *Archives of Internal Medicine*, 158, 1113–1120.

Hausdorff, J.M., Levy, B.R., & Wei, J.Y. (1999). The power of ageism on physical function of older persons: reversibility of age-related gait changes. *Journal of the American Geriatrics Society*, 47, 1346–1349.

Hess, T.M., Auman, C., Colcombe, S., & Rahhal, T. (2003). The impact of stereotype threat on age differences in memory performance. *Journal of Gerontology: Psychological Sciences*, 58B, P3–P11.

Horton, S., Baker, J., Côté, J., & Deakin, J.M. (2008). Understanding seniors' perceptions and stereotypes of aging. *Educational Gerontology*, 34, 997–1017.

Krampe, R., & Ericsson, A.E. (1996). Maintaining excellence: Deliberate practice and elite performance in young and older pianists. *Journal of Experimental Psychology: General*, 125, 331–359.

134

sean horton

Levy, B.R. (1996). Improving memory in old age through implicit self-stereotyping. *Journal of Personality and Social Psychology*, 71, 1092–1107.

Levy, B.R. (2000). Hand-writing as a reflection of aging self-stereotypes. *Journal of Geriatric Psychiatry*, 33, 81–94.

Levy, B.R., & Banaji, M.R. (2002). Implicit Ageism. In T.D. Nelson (Ed.), *Ageism: Stereotyping and prejudice against older persons* (pp. 27–48). Cambridge, MA: MIT Press.

Levy, B.R., & Langer, E.J. (1994). Aging free from negative stereotypes: Successful memory in China and among the American deaf. *Journal of Personality and Social Psychology*, 66, 989–997.

Levy, B.R., & Myers, L.M. (2004). Preventive health behaviors influenced by self-perceptions of aging. *Preventive Medicine*, 39, 625–629.

Levy, B., Hausdorff, J.M., Hencke, R., & Wei, J.Y. (2000). Reducing cardiovascular stress with positive self-stereotypes of aging. *Journal of Gerontology: Psychological Sciences*, 55B, 205–213.

Levy, B., Slade, M.D., Kunkel, S.R., & Kasl, S.V. (2002). Longevity increased by positive self-perceptions of aging. *Journal of Personality and Social Psychology*, 83, 261–270.

Lockwood, P., & Kunda, Z. (1997). Superstars and me: Predicting the impact of role models on the self. *Journal of Personality and Social Psychology*, 73, 91–103.

Lockwood, P., Chasteen, A., & Wong, C. (2005). Age and regulatory focus determine preferences for health-related role models. *Psychology and Aging*, 20, 376–389.

Lox, C.A., Martin, K.A., & Petruzzello, S.J. (2003). *The Psychology of Exercise: Integrating Theory and Practice*. Scottsdale, AZ: Holcomb Hathaway Publishers.

Maharam, L.G., Bauman, P.A., Kalman, D., Skolnik, H., & Perle, S.M. (1999). Masters athletes: Factors affecting performance. *Sports Medicine*, 28, 273–285.

Montepare, J.M., & Zebrowitz, L.A. (2002). A social-developmental view of ageism. In T.D. Nelson (Ed.), *Ageism: Stereotyping and prejudice against older persons* (pp. 77–125). Cambridge, MA: MIT Press.

Nosek, B., Banaji, M.R., & Greenwald, A. (2002). Harvesting implicit group attitudes and beliefs from a demonstration web site. *Group Dynamics*, 6, 101–115.

O'Brien Cousins, S. (2000). 'My heart couldn't take it': Older women's beliefs about exercise benefits and risks. *Journal of Gerontology: Psychological Sciences*, 55B, 283–294.

O'Brien Cousins, S. (2003). Seniors say the 'darndest' things about exercise: Quotable quotes that stimulate applied gerontology. *The Journal of Applied Gerontology*, 22, 359–378.

Ory, M., Hoffman, M.K., Hawkins, M., Sanner, B., & Mockenhaupt, R. (2003). Challenging aging stereotypes: Strategies for creating a more active society. *American Journal of Preventive Medicine*, 25, 164–171.

Payne, V.G., & Issacs, L.D. (2008). *Human Motor Development: A Lifespan Approach*. New York: McGraw-Hill.

Salthouse, T.A. (1984). Effects of age and skill in typing. *Journal of Experimental Psychology: General*, 113, 345–371.

Sarkisian, C.A., Prohaska, T.R., Wong, M.D., Hirsch, S.H., & Mangione, C.M. (2005). The relationship between expectations for aging and physical activity among older adults. *Journal of General Internal Medicine*, 20, 911–915.

Statistics Canada (2005). Canadian Community Health Survey, 2000/01. Retrieved January 9, 2006 from http://www.statcan.ca/english/freepub/82–221–XIE/00502/tables/html/2165.htm.

Steele, C.M. (1997). A threat in the air: How stereotypes shape intellectual identity and performance. *American Psychologist*, 52, 613–629.

Whitbourne, S.K., & Sneed, J.R. (2002). The paradox of well-being, identity processes, and stereotype threat: Ageism and its potential relationships to the self in later life. In T.D. Nelson (Ed.), *Ageism: Stereotyping and prejudice against older persons* (pp. 247–273). Cambridge, MA: MIT Press.

sean horton

CHAPTER NINE

MASTERS SPORT AS A STRATEGY FOR MANAGING THE AGING PROCESS

RYLEE A. DIONIGI

In contemporary western society, regular physical activity is promoted and understood as a means of maintaining one's health and independence, particularly for older people. The social and economic concerns of an aging population have prompted governments and businesses alike to provide opportunities for older people to participate in sport and exercise. For example, advocating participation in Masters sport has become part of the existing health promotion and 'successful aging' or 'aging well' discourses. In other words, older people are now encouraged to regularly participate in sport to improve their physical, mental, and social health, and, consequently, to delay the onset of age-related diseases, disability, and dependency on the health and welfare systems. The emphasis here is on self-responsibility for achieving and retaining good health, resisting the aging process, and postponing 'deep old age' (Gilleard & Higgs, 2000).

Previous qualitative research on older Masters Athletes indicates that perceived benefits to one's health and overall quality of life are key reasons why older people regularly compete in sport (Grant, 2001; Roper et al., 2003). Many older athletes themselves appear to be internalizing the health promotion message that to 'age well' means to be physically, mentally, and socially active for as long as possible. For example, 'use it or lose it' is a common catch-phrase among older people that has shown up in the context of Masters sport (Dionigi, 2008; Grant, 2001). In other words, many older people are particularly concerned about the physical, mental, and social 'losses' due to aging, so they believe that they have to find ways to continually use their body (and mind) to avoid or delay such outcomes. In essence, there is societal and individual fear tied to the loss of identity or self through frailty and dependency (Chapman, 2005; Holstein & Gubrium, 2000). Furthermore, the aging experience is commonly understood in biological terms (Phoenix & Grant, in press). Therefore, managing an aging identity appears to be of particular significance to older people who assign great importance to their able, functioning body. How do older sportspeople maintain their identity as an 'athlete' in the face of an aging body? How will older athletes cope when their body does not function as it once did? Exploring the meanings and experiences of sports participation in later life has the potential to illuminate how sport is used as a strategy for managing the aging process.

The process of negotiating the physical realities of an aging body and the psychosocial processes of an aging identity is not straightforward. This chapter highlights the complex and contradictory nature of older people's involvement in Masters sport and points to potential consequences of this behavior. To achieve these aims, I draw on observational and interview data collected from 138 athletes aged over 55 who competed at the Eighth Australian Masters Games. These data are interpreted within a framework of prominent life-stage theories, as well as traditional and postmodern understandings of identity and what it means to 'age well'. The findings reveal that Masters sport is a strategy for fighting, monitoring, adapting to, avoiding and/or accepting the aging process for these individuals. More importantly, however, the findings show that the management of an aging identity depends largely upon the individual's interpretation of what it means to 'age well'. In other words, the meanings older athletes attach to their aging experience are shaped by the broader 'aging well' discourses in which they invest. The chapter concludes by raising questions about the dominant notion of 'successful aging', or what it means to 'age well', and points to issues requiring further research.

MULTIPLE UNDERSTANDINGS OF IDENTITY, AGING, AND LATER LIFE

While there are multiple and contradictory theories of aging available, this discussion outlines prominent life-stage theories, as well as traditional and postmodern understandings of identity management in later life. This discussion is linked to an 'aging well' narrative presented by Chapman (2005), as well as research on sport and leisure in later life, to provide a framework for interpreting the motives and experiences of older Masters Athletes. Chapman compares traditional understandings of aging and identity with more recent interpretations to show how what it means to 'age well' has changed over time.

Traditionally, identity development has been about self-integration; that is, constructing a coherent sense of self or achieving ego integrity (Erikson, 1997). Therefore, aging well meant achieving 'self-integration in relation to particular sets of resources or forms of engagement' (Chapman, 2005, p. 9). In other words,

rylee a. dionigi

self-integration or reaching a developmental end-state was considered necessary to age well. According to Chapman, this key assumption has informed conceptualizations of the nature of aging well. Kleiber (1999, p. 164) argues that, 'There are few ideas about growth and adjustment in later life more compelling than Erikson's notion of establishing *ego integrity*' (emphasis in original). Erikson's (1997) nine-stage life cycle model has been influential in understanding aging from a psychosocial standpoint.

Old age, the eighth stage in Erikson's life cycle model, focuses on the developmental issue of ego integrity versus despair. To Erikson, old age involves confronting the task of integrating one's achievements and failures in the hope of finding meaning, balance, coherence, centeredness, and acceptance in life. Ego integrity refers to accepting one's life as is and not wanting to replace it with any other. Achieving ego integrity involves contemplation, adaptation, and reflection, as well as establishing a sense of connectedness and wholeness between oneself and the rest of the world. In linking this to leisure in later life, Kleiber (1999) argues that older people's participation in leisure activities that strengthen enduring parts of the self and involve interaction with others is beneficial to reaching a sense of ego and social integration. On the other hand, failure to achieve ego integrity can result in despair. Erikson (1997) argues that despair is expressed in regret, displeasure, hopelessness, uncertainty, and disgust with one's life and a feeling that life is too short to change one's path. The issues of accepting and adapting to older age, rather than fighting or ignoring it, appear central to Erikson's model. Arguably, from this perspective, the idea of older people fighting their aging body and possibly attempting to defy (or deny) old age through continued involvement in sport would be considered more detrimental to an aging identity than helpful. For instance, having a fixed sense of oneself as physically able, competitive, energetic, socially engaged, and independent may prove problematic in old age due to the eventuality of bodily decline as an outcome of increased longevity.

Other prominent aging theories that have reinforced the notion that self-integration is key to aging well are the disengagement theory, activity theory, continuity theory, and the model of successful aging (see Chapman, 2005). It is not the intention of this chapter, however, to discuss each theory in detail, but to briefly highlight how these theories have shaped assumptions about the nature of aging well. The disengagement theory (Cumming & Henry, 1961) emphasizes the withdrawal of older adults from productive social roles and society's withdrawal from them in order to maintain social equilibrium. The relinquishment of certain roles (such as employment or raising children), and the devotion of more attention to a limited number of roles within the private sphere of family and friends, were believed to be beneficial to one's aging and sense of integration (or self-centeredness; Chapman, 2005). In contrast, the activity theory (Havighurst, 1963) asserts that older people will be most fulfilled and 'integrated' if they remain active and preserve as many social roles and responsibilities as possible. This theory assumes that, for a person to age well, they should continue to do activities they have done in middle adulthood, or replace activities that must be abandoned (Biggs, 1993). 'The more active, the better' assumption underlying this theory has been criticized because the maintenance of previous levels of functioning is not always possible, and the personal meaning of the activity needs to be considered. Furthermore, the disengagement and activity theories focus on 'framing aging well in relation to the interests of society' (Chapman, 2005, p. 12), whereas the following theory — the continuity theory — emphasizes the importance of the individual.

The continuity theory posits that older people will age well if they are able to 'preserve and maintain existing psychological [internal continuity] and social patterns [external continuity] by applying familiar knowledge, skills and strategies' (Atchley, 1993, p. 5). In this theory, continuity is understood as 'a source of security and integrity in later life' (Kleiber, 1999, p. 113). In this sense, it is the maintenance of meaning and consistency of self that the activity or role and its social context hold for the individual that are most important, not the activity per se (Atchley, 1993, 1997). Past research has found continued sports participation as a key adaptive strategy for coping with changes associated with aging (Langley & Knight, 1999; Roper et al., 2003). As previously mentioned, however, the ability to maintain existing physical and mental patterns (or a sense of self-consistency) becomes difficult due to the aging body, as well as the loss of 'significant others' (see Kleiber, 1999). Furthermore, the continuity theory positions individuals as active agents who have the ability and resources (i.e., the choice) to adapt to the aging process and maintain a coherent sense of self (Chapman, 2005). The model of successful aging (Rowe & Kahn, 1998) also assumes individual agency, yet individuals were not only expected to be responsible to themselves, but also to society.

Rowe and Kahn (1998) argue that, to age successfully, individuals are to avoid disease and disease-related disability and maintain cognitive and physical function (e.g., through exercise and leisure), as well as an active, productive engagement in social life (by maintaining close relationships and involvement in personally meaningful activities) for as long as possible. Therefore, aging well was about individuals 'growing old with good health, strength, and vitality' (Rowe & Kahn, 1998, p. 23) for the 'good' of oneself and for the 'good' of society. In other words, active engagement in later life is prescribed 'to avoid the loss of self through pathological aging' (Chapman, 2005, p. 13) and prevent becoming a burden to society. On the one hand, taking on some of the responsibility for

140

rylee a. dionigi

the way in which one ages is a potentially empowering experience (Rowe & Kahn, 1998), particularly for those who have the means, ability, and desire to do so. It is often assumed in the literature that older people who compete in sport are resisting the aging body and/or feeling empowered (e.g., Coakley, 2001; Spirduso, 1995). On the other hand, such an understanding of aging well ignores individual and social determinants of health and produces a marginalizing context in which inactivity, inability, decline, and ill-health in old age are seen as representing immorality, laziness, and/or deviance (Fullagar, 2001; Jolanki, 2004). 'In such a context, guilt and shame are possible effects for those who fall ill, are dependent on others, or require health care in old age' (Dionigi & O'Flynn, 2007, p. 373). The assumptions tied to the model of successful aging are consistent with the key messages underlying the theory of the Third Age; therefore, the same criticisms apply to it.

Traditionally, old age was understood as the final stage of life, usually commencing after retirement, or at age 65. More recently there has been a rethinking of later life into two sequential stages: the Third Age and the Fourth Age (Laslett, 1989, 1996). The Third Age is characterized by 'relative' freedom, health, leisure, personal achievement, and independence. This phase of life generally begins at retirement from work or family obligation, is expected to last about 30 years, and should be devoted to self-fulfillment through various activities (Laslett, 1989). According to Laslett, aging well is about individuals continuing physical, social, and mental activities that provide pleasure and self-worth so 'that the Fourth Age will come as late and be as brief as possible' (1989, p. 61). In this sense, old age (or the aging body) is a state to be resisted or postponed, not contemplated or accepted as articulated by Erikson. Consequently, life in the Third Age can be accompanied by repression, avoidance, and denial of the Fourth Age (Blaikie, 1999). The Fourth Age or 'deep old age' is characterized by sickness, dependency, decrepitude, frailty, and the imminence of death. And although one can enter this stage at any time, it is usually compressed to the last couple of years before death (Blaikie, 1999; Laslett, 1989). Furthermore, key dimensions of the Third Age (and hence, 'aging well') are individual agency (or personal empowerment), flexibility, and choice, which are also characteristic of postmodern interpretations of the management of an aging identity.

From a postmodern perspective, identity is interpreted as something which is shifting not fixed, chosen not given, multiple not singular, and open-ended not integrated (see Biggs, 1997; Murphy & Longino, 1997; Sarup, 1996). Postmodernity is characterized by plurality, which presents individuals with a variety of identity choices or the opportunity to construct and reconstruct multiple selves. In this sense, aging well has become understood as the open-ended, meaningful negotiation of multiple selves among later life resources, events, and changes

141

(Chapman, 2005). The diversity of options available to individuals in postmodern society means that older people can take on new roles or activities and create alternative identities in later life. For example, someone who did not play sport as a child may begin competing in sport in later life and take on an identity as an athlete. Alternatively, a person who has been athletic all their life may identify themselves as a passionate reader and spectator of sports later in life. It also means that older people who may have limited resources or physical and mental abilities can still be involved in a 'meaningful negotiation of selfhood' (Chapman 2005, p. 15). That is, regardless of one's level of health or wealth, individuals may still experience ongoing contentment and a sense of coping in old age, without the need to reach integration. If aging well is understood in this way, older athletes who can no longer use their body or compete in sport as they once did may cope with aging and maintain a sense of self in old age. As summarized by Bauman, in the past, identity management 'was about how to construct an identity and keep it solid and stable, [whereas] the postmodern "problem of identity" is primarily how to avoid fixation and keep the options open' (Bauman 1995, p. 81).

The multiple, shifting, and (at times) conflicting understandings of aging, identity, and what it means to age well, highlight the importance of understanding how older people perceive and make choices relative to their personal and social contexts. More specifically, how does the choice to compete in sport intersect with these multiple understandings and assist in the management of an aging identity? The above discussion highlights that the 'use or lose it' mantra (which is common among the reasons why older people participate in sport) is also strongly communicated through the activity theory, continuity theory, successful aging model, and the theory of the Third Age. Are older sportspeople investing in these understandings of aging and identity when they explain why they compete in sport and when they actually participate? Are they demonstrating a resistance or adaptation to aging and/or are they acting out a fear or acceptance of old age? How does the take-up of different understandings of what it means to 'age well' affect the management of identity in later life? What are the potential individual and societal effects of these different interpretations or discourses of 'successful aging'? To address these questions, I draw on observational and interview data from a major study of older Masters Athletes (see Dionigi, 2008).

THE COMPLEX AND CONTRADICTORY NATURE OF OLDER PEOPLE'S INVOLVEMENT IN MASTERS SPORT

To interpret the meanings and experiences of sports participation in later life within the above framework, I present examples of the talk and action of 138

rylee a. dionigi

athletes (70 women and 68 men, aged 55–94 years) who competed at the Eighth Australian Masters Games. The participants were white, English speaking, and middle class, which is representative of the 'typical' demographic of athletes competing in Masters sports in Australia. These people were regularly involved in physically demanding individual events (e.g., track and field, road running, swimming, cycling, gymnastics, triathlon, tennis) and team sports (e.g., soccer, basketball, ice hockey, field hockey, beach volleyball, badminton). Half of the sample were 'sport continuers' (i.e., remained competing since a young age), while a quarter had played in their youth and restarted later in life ('rekindlers'), and the remainder were 'late starters' to sport (i.e., commenced playing sport in their 50s or older). Two broad themes to emerge from this research in relation to their meanings, motives, and experiences of Masters sport were 'I'm out here and I can do this!' and 'use it or lose it.' A discussion of these themes and their various sub-themes reveals how Masters sport was a key strategy for managing the aging process for these individuals.

'I'm out here and I can do this!'

As was mentioned in the opening to this chapter, managing an aging identity has significance to older people who assign great importance to their able, functioning body due to biological aging. Competing in Masters sport provided participants with confirmation that they are coping with the aging process, retaining control over the use of their body, and, therefore, have not yet entered the Fourth Age. Sports participation involves physical exertion, agility, strength, fitness, skill, and competition. Therefore, competing in sport provided participants with 'the satisfaction of knowing that I'm not losing it' (62-year-old male 300-meter hurdler). In other words, it provided them with the reassurance that 'I can still do it! I'm not too old' (60-year-old female softball player). That is, participants experienced a sense of pride, achievement, empowerment, and reassurance from the knowledge (and practical demonstration) that they had not lost their physical ability and were still capable of competing in sport, regardless of when they began sports participation.

When asked why they compete in sport, many participants, especially the sport 'continuers' and 'rekindlers', said that they were from a 'sporty family' or had always been a 'physical person' who valued competition, sport, and fitness. For example, a 65-year-old female squash player said, 'I think it's just *me*. I've done it for so long . . . I probably started the competition sport when I was about 13 [years] . . . and I've continued all the way through . . . and I enjoy doing it.' A male basketball player, aged 62, explained, 'We still play tough, we get out there and put our all in, and I play with a competitive spirit, which is very strong. So we hate to lose . . . we love to win'. For some participants, competing in sport was their 'lifeblood', as a 57-year-old male baseball player explained, 'Any sport, once it's in your blood, it's in your blood and you just love it'. Furthermore, a 60-year-old past Australian representative basketball player said that involvement in competitive sport was:

> just my lifestyle. What . . . I've done all my life. I've been involved in sport from day one, which means it's a natural partnership of mine, like some people . . . might read a book, but I can't do that. I've got to be active and it's just that continuation.

It was not only sport continuers, but also 'late starters' to sport who defined themselves as competitive. Many of these people said they were (or are simultaneously) competitive in other contexts, such as employment, which perhaps indicates that they now express their competitiveness in a sporting context. For example, an 82-year-old female track and field athlete who competed in her first Fun Run at age 60 said:

> Competitiveness I suppose that's . . . one thing that keeps you going too, because you're always trying to better something . . . If anyone's against me I think, Well, I've got to beat them . . . I mean, I'll *try* to beat them, not got to, but I'll do . . . my best to sort of beat them. It's still there, I've got that little bit of *streak* still there [laughs] . . . I suppose it's just the competitiveness in me is why I do it anyway. I can't see that there'd be anything else. It's just that *drive* I suppose that's there, where perhaps another person, just wouldn't do it . . . I suppose my nature makes me keep pushing.

These findings demonstrate that both men and women of varying ages and sporting histories, and participants of both individual and team sports, identified themselves as sportspeople or athletes, and they were expressing this identity through their continued involvement in Masters sport. The participants' use of the words and phrases, such as 'my nature', 'it's just me', 'driven', a 'streak still there', and 'it's in your blood', are consistent with traditional notions of identity as self-integration or 'the real me' (Erikson, 1968, p. 19) and the continuity theory (Atchley, 1993, 1997).

The participants appear to be attempting to maintain an athletic identity by proving to themselves (and others) that they are out there competing and they are still capable of effectively using their bodies. For example, a 71-year-old man, who had competed in gymnastics in his youth but did not return to the sport until he was aged 50, said:

144

I just like to be able to . . . think that I can still do something. That I am still capable of throwing my body around and pulling it and twisting it, turning it and being able to put it where I want to put it and, of course, keep fit at the same time.

Likewise, a 73-year-old woman who ran in her first half marathon at age 67 explained:

Well, it is just satisfying to know you can run a half marathon or swim 1,500 meters, which I do, and that makes you feel as though you are actually still here.

From the perspective of a 'sport continuer', a 68-year-old male baseball player said:

I remember when I was in my 40s, I thought, 'Well, I'll try and make 50', and when I got to about 49, I said, 'All right, 55, that's a *good*, good age'. Then when I got to 55 I thought, 'maybe 60?' . . . and . . . somewhere down in there, there's a little bit of ego, 'Hey man, yeah, I'm still out here, I wonder how long I can keep this up?' It's sort of like a king of the mountain . . . [It provides] confirmation that things still work, legs and arms still work . . . I go out and I run and my knees don't break and my hamstring doesn't tear, and I pitch and I still am able to throw a curve ball, yeah, it still works!

Masters sport provides a context in which older people can test, push, and monitor the competency of their aging bodies. This practice provided participants with verification of their healthy, physically active, and athletic identity, which contributed significantly to feelings of empowerment and control.

These feelings of strength, confidence, and competency generalized into an overall perceived sense of control and independence, which was encapsulated in the common phrase, 'I can do everything I want to'. For instance, many participants spoke of the general benefits that they associated with competing in sport:

By keeping fit, you keep mentally alert and you are able to do things . . . I think it's a benefit all round. I ride and my wife runs . . . So I think it [competing in sport] increases our options to enjoy our lifestyle. We can do more things, more energy.

(71-year-old male cyclist)

145

> You're setting your own quality of life by being fit and active. I mean, if I didn't do anything, I mean, I probably couldn't mow my lawn even. I'd have to get someone to do that, because . . . I wouldn't be fit enough to do it . . . It keeps your mind active . . . you can react to things a bit quicker.
> (60-year-old female squash player)

The above quotes show that the participants perceived 'aging well' as making the choice and having the ability to maintain a healthy, physically active, and socially engaged lifestyle. The participants' words and actions typify understandings of life in the Third Age (Laslett, 1996) and are consistent with the key assumptions underlying the continuity theory and the model of successful aging. They were experiencing a sense of self-fulfillment and personal empowerment through using their bodies in physical, competitive, and powerful ways. They were also experiencing a sense of independence and control over their life that extended beyond the context of sport.

These findings indicate that taking on some responsibility for the way in which one ages can be a personally empowering experience for many, as argued by Rowe and Kahn (1998). The above feelings and experiences expressed by participants indicated to themselves (and the general public) that they were not (yet) experiencing ill health, disability, isolation, or dependency in older age. Continued participation in Masters sport was a strategy for monitoring and adapting to the aging process, as well as a way to maintain a coherent sense of self as physically active, capable, competitive, and productive. To retain these feelings, abilities, and sense of self, however, participants believed that they had to use their body (and mind) as much and for as long as possible.

'Use it or lose it'

This theme provides insight into how the 'use it or lose it' motto operates as meaningful within the context of Masters sport participation among older people. Study participants believed that if they stopped participating in sport, they would soon become 'old', 'age badly', 'rust up', 'end up in a nursing home', or enter what Laslett (1989) calls the Fourth Age. For example, a 65-year-old male badminton player explained the necessity of sport in his life: 'You have to play [a sport]. If you stop, you seize up. That's the problem once you get to our age. You have to keep moving'. A female tennis player elaborated:

> It's a way of life when you get to 81 like me . . . People kind of get to 60 and think life's had it, but as long as you keep fit, eat right, exercise [grins].

146

rylee a. dionigi

It's a way of life . . . to wear out not rust out [laughs] . . . You just have to keep going and keep on keeping on.

A 71-year-old female badminton player agreed:

Well, I'll tell you what [playing badminton is] all about. If you don't use it, you lose it. It's not mine, it's a well known one . . . You have really got to keep your body going, because if you don't, you just rust up [she says matter-of-factly]. You just rust up like anything else.

A 70-year-old female swimmer acknowledged that it was not only about using the body:

I think keeping active — use it or lose it [she smiles] . . . Use your body and your mind, you know, keep your mind active. Do things that involve thinking and concentration, and also use your body in doing physical things [to maintain] . . . mobility, flexibility, that's physically and mentally [laughs] . . . they just become stagnated, I suppose.

The concern of potential institutionalization in old age was also motivating participants to maintain involvement in Masters sport:

I just suppose, the thing is you don't want to get old. You want to keep moving, keep mobile, active [and] . . . playing sport against [younger people] . . . Old is when I can't move around properly, I suppose [she chuckles]. I don't want to be one of those persons, like you see in a nursing home that are just — [she demonstrates what she means by sitting limp with her head down].

(66-year-old female netball player)

Male participants of varying ages and sports also recognized the importance of retaining one's health in later life:

If you don't use . . . your body as a whole, you know, all the components in your body, your heart, your lungs, your liver, your kidneys, the lot, and externally your muscles . . . well, you're losing, I suppose, the best part of your life really . . . If your muscles don't work and you don't exercise, your knees start to go, arthritis sets in, and it all builds up. You've got to keep things moving. That's why I do a lot of flexibility exercises to keep my joints free.

(85-year-old male runner)

Unless I do my exercises, unless I go training, I become very old very quickly, as well as very fat [he grins] . . . for my health [competing in and training for cycling] is critical.

(65-year-old male cyclist)

Well, if I wasn't healthy, I wouldn't be able to do *any* of the things that I do . . . I couldn't do . . . all the things that go with *living* . . . The body can only last that long, certain organs are going to start to *fail* eventually. By exercise, it prolongs your organs functioning properly . . . prolong[s] . . . your ability to *move* and then do things.

(65-year-old male beach volleyball player)

Although the participants recognized the health 'risks' associated with older age, they expressed and acted out a resistance to the aging process by continued participation in, and regularly training for, Masters sport. These practices were an almost desperate attempt to fight, avoid, or delay losing their physical ability and health, because with these losses they believed they would also lose their independence, a sense of control over their life, and a sense of self. In other words, the feeling of personal empowerment experienced through Masters sport participation was primarily driven by a fear of its opposites — loss of control, loss of ability, loss of health, and loss of selfhood.

Therefore, when 'aging well' is understood as self-responsibility for health, continuity of previous levels of activity, maintaining a core self and reaching ego integrity, the tenuous balance between a sense of personal empowerment and the desire to fight the aging process is illuminated. Clearly, the participants are very much trying to avoid entering the Fourth Age. Laslett argued that the onset and duration of the Fourth Age 'should be put off for as long as possible by appropriate behavior in the Third Age [that is, continued activity of body and mind in the pursuit of self-fulfillment]' (1989, p. 154). The participants believed that the more active and intense they were, the better it was for their health. In doing so, the participants have adopted the assumptions implicit in the theory of the Third Age, continuity theory, activity theory, and model of successful aging. However, these theories (which promote this 'keep going' mentality) have been criticized as 'escapist' responses to the realities of the physiological aging process and as rather idealistic (Gilleard & Higgs, 2002). Consequently, what these findings highlight are the individual and cultural fears tied to the potential health risks of aging (Chapman, 2005) and the potential for individuals to find ways to avoid or deny old age.

Several participants said that they 'turn a blind eye' to the physical realities of the aging process, indicating that Masters sport was a strategy to ignore the

148

eventuality of deep old age. For instance, a 60-year-old female squash player, who does a minimum of four aerobics classes per week, swims and/or runs once a week, and plays squash twice a week, said the following when asked how long she intended to keep up this schedule:

> As long as I can . . . I think now you've got this mind-set that you have to keep your body . . . moving and clean . . . I probably don't look . . . that far ahead. While I'm able to do it, I don't want to look at the sort of negative side . . . so while I'm able to [do it] . . . I turn a blind eye to maybe what's down the track a bit.

An 85-year-old male, who regularly competes in Fun Runs, swims in his backyard pool, and works out in his home-based gym, responded in the following way when asked how he would cope if he could not run anymore:

> I don't know. I don't want to think about that too much, [slight laugh] . . . I just want to keep going as long as I can . . . I will obviously slow down a lot, but I'm determined to keep going, and the current rate of decline is so minimal that it appears that if I can keep up this [fitness] program, I'll be all right for another five years or so, and still run about the same level as I am now, which would be great.

While some participants did not like to think about possible 'risks' or 'losses' associated with aging, others perceived Masters sport participation as a means to avoid old age altogether. A 60-year-old male badminton player explained, 'Playing badminton keeps me alive! . . . I've always been a competitor, and I probably will be when I die . . . If I die on the badminton court I'd be happy'. Due to the physically intense and at times high-risk nature of sports participation in later life, it is possible that some people may die in the field of play and avoid deep old age. When discussing his disdain for aged-care homes, a 66-year-old gymnast said, 'Well, that's the point, what we are doing here [competing in gymnastics], we will most probably just go "plop" one day, and that will be a good thing . . . fall from a great height on our head'. Furthermore, a 76-year-old female tennis player said that she would rather jump to her death than be institutionalized:

> I have a friend I see . . . who's in a nursing home . . . This woman is 85, and I know through her what it is like to be in a home and I say to my husband, 'Where is the nearest *gap*? We shall jump over it', rather than do that. It's like an imprisonment. She feels it. It's morbid, it's sad, it's unfeeling.

If people do not die suddenly, however, old age will eventually come to them. Therefore, attempting to ignore or avoid issues associated with aging may be maladaptive to an aging identity. For instance, if aging well is understood as reaching an integrated end-state and accepting oneself as 'old' (Erikson, 1997), then older people attempting to avoid or ignore thoughts of existential issues, such as reflecting on the meanings of one's life and coping with bodily limitations or the nearness of death, may never reach ego integrity (Biggs, 1997, 1999).

On the other hand, several participants recognized the eventuality of deep old age and indicated an acceptance of the aging process. As one woman said, '. . . it's another two years and I'm 70 . . . you can't keep going forever, and I won't be able to play hockey — this might be my last [year] . . . but then that doesn't matter, been there, done that . . . *It's life!* You just take it all as it comes'. An 81-year-old male cyclist said:

> Well, I think that I'm mature enough to know that, ok, this is something that happens, it's not my doing, but while it is left up to me, I will do it. When JC [Jesus Christ] says to me that, 'Listen son, your body's worn out [slight laugh], you can no longer do that', well, then my body will tell me, but it will not be my desire to do it . . . I've got enough common sense to know that others have fallen by the wayside and look younger than me . . .

A 76-year-old female tennis player elaborated on what it means to her to be competing in sport at this point in her life:

> Making these latter years as enjoyable as I can by being fit . . . You *can't avoid* old age, it's inevitable [slight pause]. I count myself lucky, for the things I've done and the way I feel now. I've just had all the check-ups, and I seem to be going all right, and I think that's why sport and still going with it at my age is . . . important . . . I just can't believe that I haven't got that same get-up-and-go, and I've sort of said, 'Well, if this is it, the way you are going to go, you've got to get used to it, but aren't you lucky that you've been so fit for so long?'

A 68-year-old male baseball player explained:

> I suppose it's just a very balanced sort of recipe, and at some point, something will become overpowering and you say, 'Ok, that's it' . . . that's how everything in life should be treated. Nothing should be so obsessive that you can't just say, 'Hey, yeah, that was cool'. Even life itself, that was

rylee a. dionigi

good, I had a good run . . . So life's that way. I think baseball's that way, I think everything you do in life should be that way. You do it as long as you enjoy it and you gracefully back out . . .

These participants accepted that they will not always be able to continue Masters sport participation, so therefore they are making the most of the 'here and now' while they have the ability, means, and desire to do so. In other words, their current involvement in Masters sport represented lives well lived and lives lived to their fullest. This is not to say the participants did not have other leisure or social interests, but the strenuous physical activity associated with Masters sport was the practice that they believed helped most in keeping them 'socially, mentally, and physically alive'. As a 79-year-old male said, 'I always look at the positive side . . . maybe I've had my life, but I've still got a lot to live . . . and [competing in Masters swimming] is the one way I'm doing it'.

Erikson's (1997) notion of ego integrity is important in making sense of this feeling of acceptance and contentment in later life. The above participants were not expressing despair (i.e., regret, discontent, and disgust with one's life, and a belief that life is too short to change one's path), but were reflecting on their accomplishments and perhaps harmonizing the tension between integrity and despair. From a postmodern standpoint, it could be argued that participants are involved in an open-ended negotiation of meaning-making that demonstrates an adaptation to the aging process and later life changes (Chapman, 2005; Phoenix & Sparkes, in press). As a 70-year-old female runner said, 'Keep battling on . . . you have just got to face life and make the most of it'.

MASTERS SPORT, 'AGING WELL', AND IDENTITY MANAGEMENT

When these older people explained why they compete, and when they actually participated in physically intense Masters sports, they typified life in the Third Age, were managing the physical realities of their aging bodies, and were negotiating the psychosocial processes of an aging identity. In essence, older people who participate in Masters sport are competing to maintain a physically active, healthy, competitive, and socially engaged life, and to delay (or avoid) ill health, disability, loneliness, and dependency in old age. However, this process is not straightforward. Masters sport was a strategy for monitoring, adapting to, fighting, avoiding, and/or accepting the aging process for these individuals. Therefore, the findings simultaneously exposed stories of personal victories and (perhaps) private desperation. They pointed to the perceived benefits and the potential consequences of this somewhat paradoxical behavior for negotiating an aging identity. The management of an aging identity depended primarily upon how 'aging well' was understood by the participants. This outcome points to the possible cultural and individual effects of the dominant discourse of 'successful aging' (or what it means to 'age well') that currently operates in society. It also highlights issues that require further research.

The participants in this study appeared to internalize Rowe and Kahn's (1998) model of successful aging. This model is the most 'popular' or dominant cultural understanding of aging well in contemporary society. When successful aging is understood in terms of the maintenance of a healthy body and an active and engaged lifestyle, the onus is on the individual for retaining good health. The older people described in this chapter show how taking action to maintain ability, independence, and health was an empowering experience. However, such an understanding of aging well is potentially problematic for those who fall ill, are disabled, lack financial resources or education, and/or are from different cultural backgrounds (Blaikie, 1999). This successful aging discourse can also contribute to a heightened undesirability, fear, or denial of ill health in old age (Gilleard & Higgs, 2002), which can encourage older people to find ways to resist or avoid deep old age. In other words, living in a culture that places a great deal of value on activity, health, competition, independence, and bodily performance can make older people feel that they 'have to' or 'should' keep active to age well, if they internalize these dominant ideals. Therefore, advocating 'continuity' of an active, social life can become maladaptive to an aging identity if it represents a denial of existential issues that may need to be faced in later life (e.g., self-reflection on the meaning of life, bodily limitations, ill health, and death; Biggs, 1997). Such understandings of aging well also have the potential to make older people who cannot or do not want to keep active feel guilty, shameful, unsuccessful, or valueless.

In addition, the participants' talk was consistent with traditional understandings of identity as self-integration, which highlighted the struggle between acceptance and denial in later life (Erikson, 1968, 1997). The participants identified them-selves as competitive and/or physically active, and they used Masters sport participation as a strategy for maintaining this sense of self. Continuity of self can be beneficial to an aging identity as it provides feelings of comfort, consistency, and self-worth (Kleiber, 1999). The findings in this chapter showed that continued involvement in Masters sport was assisting individuals to monitor and adapt to the aging process. However, identifying oneself as athletic and assigning great importance to bodily competence in later life is potentially problematic due to the unpredictability of the aging process and the eventuality of physical decline in old age. Several participants in this research indicated

152

that they will find it difficult to cope when they can no longer compete in Masters sport or maintain their physical ability and independence. When aging well is understood in terms of consistency of self and reaching a developmental end-state, older athletes who define successful aging primarily in physical terms may not find acceptance in old age. Hence, interpreting aging well in terms of self-responsibility for health, continuity, and self-integration can be just as problematic as it is helpful to identity management in later life.

The above discussion highlights the need to rethink popular notions of what it means to age well and consider alternative interpretations of identity management that could allow for an ongoing, meaningful aging experience. The postmodern shift toward understanding aging well as 'the negotiation of the co-construction and reconstruction of multiple selves in an ongoing, open-ended process of meaning–making amid later-life events and transitions' (Chapman, 2005, p. 9) has implications for further research in the area of Masters sports participation. When aging well is understood as a continual process of deconstruction and reconstruction, it is recognized that people attach multiple and contradictory meanings to their aging experience. Therefore, the focus shifts from the concerns of physiological aging to the importance of biographical aging. Identity management also moves from developing a fixed sense of self and the goal of self-integration to managing multiple selves and open-ended negotiation (Chapman, 2005). Therefore, from a postmodern standpoint, individuals can age well despite changes to health and resources because these transitions can be interpreted as part of one's life story (see Phoenix & Sparkes, 2009).

In this sense, even though older athletes may get to a stage where they can no longer participate in Masters sport, the selves associated with being an athlete in later life remain meaningful as they are rebuilt or adapted to suit their current circumstances. If the management of aging by older athletes is interpreted in terms of meaning–making, then a former athlete who becomes frail or has diminishing resources may still be hopeful and coping in later life. Their life story may demonstrate positive self-aging amid personal, physical, economic, and socio-cultural constraints. Therefore, conducting life history and narrative research on the experiences of former older Masters Athletes (i.e., individuals who no longer desire, or are unable, to compete) within a postmodern identity framework will provide insight into how they are managing in later life. For instance, to what extent do older people feel a sense of guilt, shame, failure, and/or worthlessness because of their current circumstances? And to what extent do they remain feeling empowered, hopeful, and worthy in later life? A deeper understanding of what discourses of aging well people invest in, and what alternative discourses could be made available, will allow a more meaningful 'deep old age' experience to be imagined and offer enriched insight into older

153

people's concerns and desires. Such research would be valuable for health promotion experts, researchers of sport and aging, and older athletes themselves.

REFERENCES

Atchley, R.C. (1993). Continuity theory and the evolution of activity in later adulthood. In J.R. Kelly (Ed.), *Activity and aging: Staying involved in later life* (pp. 5–16). Newbury Park, CA: Sage Publications.

Atchley, R.C. (1997). *Social forces and ageing: An introduction the social gerontology* (8th ed.). Belmont, CA: Wadsworth.

Bauman, Z. (1995). *Life in fragments: Essays in postmodern morality*. Oxford: Blackwell.

Biggs, S. (1993). *Understanding ageing: Images, attitudes and professional practice*. Buckingham: Open University Press.

Biggs, S. (1997). Choosing not to be old? Masks, bodies and identity management in later life. *Aging and Society*, 17, 553–570.

Biggs, S. (1999). *The mature imagination: Dynamics of identity in midlife and beyond*. Buckingham: Open University Press.

Blaikie, A. (1999). *Aging and popular culture*. Cambridge: Cambridge University Press.

Chapman, S.A. (2005). Theorizing about aging well: Constructing a narrative. *Canadian Journal on Aging*, 24, 9–18.

Coakley, J.J. (2001). *Sport in society: Issues and controversies* (7th ed.). St. Louis: Mosby-Year Book.

Cumming, E., & Henry, W.F. (1961). *Growing old: The process of disengagement*. New York: Basic Books.

Dionigi, R.A. (2008). *Competing for life: Older people, sport and ageing*. Saarbrüecken: VDM Verlag Dr. Müller.

Dionigi, R.A., & O'Flynn, G. (2007). Performance discourses and old age: What does it mean to be an older athlete? *Sociology of Sport Journal*, 24, 359–377.

Erikson, E. (1968). *Identity: Youth and crisis*. New York: Norton

Erikson, E. (1997). *The life cycle completed: Extended version*. New York: Norton.

Fullagar, S. (2001). Governing the healthy body: discourses of leisure and lifestyle within Australian health policy. *Health*, 6, 69–84.

Gilleard, C., & Higgs, P. (2000). *Cultures of aging: Self, citizen and the body*. Harlow: Prentice Hall.

Gilleard, C., & Higgs, P. (2002). The Third Age: class, cohort or generation? *Ageing and Society*, 22, 369–382.

Grant, B.C. (2001). 'You're never too old': Beliefs about physical activity and playing sport in later life. *Aging and Society*, 21, 777–798.

Havighurst, R.J. (1963). *Successful ageing*. In R. Williams, C. Tibbits, & W. Donahue (Eds.), *Processes of aging* (Vol. 1). New York: Atherton Press.

Holstein, J.A., & Gubrium, J.F. (2000). *The self we live by: Narrative identity in a postmodern world*. New York: Oxford University Press.

Jolanki, O. (2004). Moral argumentation in talk about health and old age. *Health: An Interdisciplinary Journal for the Social Study of Health, Illness and Medicine*, 8, 483–503.

154

Kleiber, D.A. (1999). *Leisure experience and human development: A dialectical approach*. New York: Basic Books.

Langley, D.J., & Knight, S.M. (1999). Continuity in sport participation as an adaptive strategy in the aging process: A lifespan narrative. *Journal of Aging and Physical Activity, 7,* 32–54.

Laslett, P. (1989). *A fresh map of life: The emergence of the Third Age*. London: George Weidenfeld and Nicholson Ltd.

Laslett, P. (1996). *A fresh map of life* (2nd ed.). Hampshire: Macmillan.

Murphy, J.W., & Longino, C.F. (1997). Toward a postmodern understanding of aging and identity. *Journal of Aging and Identity, 2,* 81–91.

Phoenix, C., & Grant, B. (in press). Expanding the research agenda on the physically active aging body. *Journal of Aging and Physical Activity*.

Phoenix, C., & Sparkes, A. (2009). Being Fred: Big stories, small stories and the accomplishment of a positive aging identity. *Qualitative Research*.

Roper, E.A., Molnar, D.J., & Wrisberg, C.A. (2003). No "old fool": 88 years old and still running. *Journal of Aging and Physical Activity, 11,* 370–387.

Rowe, J., & Kahn, R. (1998). *Successful aging*. New York: Pantheon Books.

Sarup, M. (1996). *Identity, culture and the postmodern world*. Athens, GA: The University of Georgia Press.

Spirduso, W. (1995). *Physical dimensions of aging*. Champaign, IL: Human Kinetics.

SECTION FOUR

TOWARD A COMPREHENSIVE MODEL OF LIFESPAN PHYSICAL ACTIVITY, HEALTH, AND PERFORMANCE

CHAPTER TEN

PHYSICAL ACTIVITY
What role does it play in achieving successful aging?

PATRICIA WEIR

The world around us is aging, and it is aging rapidly. The 2006 Census in Canada reported an 11.5 per cent increase in the number of people aged 65 years and older compared to 2001, topping the four million mark for the first time in census history. This trend is expected to continue, and will accelerate in 2011 when the first baby boomers reach 65 years of age. Seniors in Canada now account for 13.7 per cent of the total population. As the population ages, it will become increasingly important for seniors to maintain high levels of health and functional independence in order to live independently as long as possible. One framework to consider this within is successful aging (SA), also referred to as healthy aging, productive aging, and aging well, which proposes that aging does not have to lead to a negative cascade of failing physical and cognitive function.

The term successful aging (SA) is not new; it dates back over 50 years (Baker, 1958; Pressey & Simcoe, 1950). This emphasis on aging well was wholly consistent with the World Health Organization's definition of health as a state of complete physical, mental, and social well-being, rather than simply an absence of disease (World Health Organization, 1948). While there has been a great deal of attention given to the concept of SA, it was not until the seminal article by Rowe and Kahn in 1987 that the term 'successful aging' became mainstream, and provided a new way of thinking about the aging process.

While a majority of seniors identify themselves as aging successfully (Tate et al., 2003), and report their health status as being good or very good, more than one in four seniors face restrictions in their activities due to long-term health problems. Disturbingly, more than four out of five seniors living at home suffer from a chronic health condition (Health Canada, 2002). Many of these health-related problems are due to lifestyle factors, one of which is a lack of physical activity. Across the aging lifespan, the level of physical activity ranges from the sedentary adult to the recreational and/or competitive Masters Athlete (MAs). MAs are individuals who continue to train for and compete in athletic competitions. They range in age from 22 years through to 100+ years of age, and thus represent the entire spectrum of aging. Several researchers have suggested that they represent the ideal model of aging, given their higher levels of physical health (Cooper et al., 2007; Hawkins et al., 2003). However, physical health is only part of the story when it comes to successful aging, and a more comprehensive exploration of the positive health benefits is warranted. This chapter discusses current issues in defining SA, describes the multidimensional model proposed by Rowe and Kahn, and lastly examines the contribution that physical activity can play in achieving successful aging. In this final section, I explore how MAs can provide additional support for the positive benefits associated with sustained involvement in physical activity.

While SA is taking on increasing importance as the population becomes older, a lingering problem 20 years after Rowe and Kahn's (1987) paper is a lack of consensus on what defines a 'successful ager'. As Baltes and Baltes (1990) point out, 'Defining the nature of success is elusive . . . consensus about the definition of success is difficult to achieve' (p. 4). Some of the contributing factors to reaching a consensus on a definition include: (a) the different perspectives on SA, (b) the difference between objective and subjective definitions, and (c) using SA as constituent versus predictive measure.

In terms of the different perspectives on SA, it has been broadly defined in two ways. The first is that SA is a state of being that can be objectively measured at a certain moment, and the second is that SA is a process of continuous adaptation. When viewed as a state of being, SA has been defined by outcome measures related to the compression of disease and disability (Berkman et al., 1993; Fries, 1980; Rowe & Kahn, 1987), cognitive performance (Reed et al., 1998; Salthouse, 1991; Simonton, 1988), achievement in physical domains (Ericsson & Charness, 1994; Schulz & Curnow, 1988; Strawbridge et al., 1996), social support (Pennix et al., 1997; Seeman et al., 1995), life satisfaction (Palmore, 1979), and mastery/growth (Baltes & Baltes, 1990; Schulz & Heckhausen, 1996). As Schulz and Heckhausen (1996) suggest, these measurable domains of functioning can be applied to any stage of life, and they all subscribe to the criteria that the higher the level of functioning, the more successful the individual. The second approach views SA as a process of continuous adaptation. This view suggests that individuals accept age-related decrements in function and performance and do the best they can with what they have (Baltes & Baltes, 1990). Both of these views have been encompassed in multicriterion models that rely on a number of factors and variables to define SA.

In a recent review of quantitative studies on SA, Depp and Jeste (2006) identified 29 different definitions from 28 studies. On average there were 2.6 components per definition with a range of one to six components. The majority of the components identified could be broadly classified as being either biomedical or psychosocial in nature. Biomedical theories focus on the optimization of life expectancy and the minimization of mental and physical disability, typically reflected through measures of cognitive and physical functioning (Bowling & Dieppe, 2005). In contrast, psychosocial theories emphasize life satisfaction, personal growth, social engagement and participation, a sense of control, self-efficacy, and self-worth.

The majority of research studies to date have relied on objective definitions related to the researcher's interest. This is clearly highlighted in the extensive reviews on defining SA published over the last couple of years (Bowling & Illife, 2006; Depp & Jeste, 2006; Peel et al., 2004). According to Depp and Jeste (2006), 26 of the 29 definitions included disability and/or physical functioning, evaluated through measures of self-reported activities of daily living (ADLs), and physical performance measures such as grip strength, lifting weights, and climbing stairs. In contrast, only 13 definitions included a cognitive component assessed through self-reported memory functioning, sentence completion tasks, and cognitive screening tests. While objective measures have allowed a wide array of measures to be examined within the context of SA, they do not take into account the individual perceptions of older adults. Subjective reports of perceptions of SA (Bowling & Illife, 2006; Strawbridge et al., 2002), focus groups (Reichstadt et al., 2007), surveys (Fisher, 1995; Phelan et al., 2004; Tate et al., 2003), and semi-structured interviews (Knight & Ricciardelli, 2003) have been used to elicit information on how older adults view the process of aging, and what components of SA they deem most relevant and important. Fisher reported that SA was described relative to an individual's orientation to the present and future, and included the constructs of activity, income, health, interactions with others, and a positive attitude. Knight and Ricciardelli (2003) interviewed older adults and asked them, 'What do you think successful aging is?' All respondents provided at least one response to the first question, with the majority of adults providing one or two responses. Overall, a total of 164 responses were given, comprising eight themes. Health was the most frequently occurring theme (mentioned by 53.3 per cent of participants), followed by activity (50 per cent of participants). One respondent stated that successful aging was 'primarily keeping fit. Your health is everything . . . other things come along and you can't accept them if you're not healthy' (Knight & Ricciardelli, 2003, p. 228). Following the interview, participants were asked to rate the importance they would place on each of ten criteria for SA. The final ranked order of the criteria from most to least important was: health, happiness, mental capacity, adjustment, life satisfaction, physical activity, close personal relationships, social activity, sense of purpose, and withdrawal. In terms of their beliefs about aging, most participants accepted their aging and were happy with their current age. They also believed that their aging was better than they expected, and that 'old' was an attitude more than a chronological age.

Tate et al. (2003), in a follow-up study to the Aging in Manitoba Longitudinal Study, asked a group of retired Air Force aircrew, 'What is your definition of successful aging?' and 'Would you say you have aged successfully?' Overall, 83.8 per cent of the respondents felt they were aging successfully. In terms of their definition of SA, all responses were reduced to 20 component themes. Again, the most frequent component, identified by 30 per cent of the respondents, was 'health'. This was followed by satisfaction, keeping active (general), and keeping physically active. The top ten themes are summarized in Table 10.1. Reichstadt et al. (2007) asked members of 12 focus groups two questions: (a) How would you define successful aging? and (b) What are the necessary components of SA? Participants identified 33 categories that fit into four major themes: health/wellness, engagement/stimulation, security/stability, and attitude/adaptation.

Phelan et al. (2004) used a different methodology to assess attributes that characterize SA. They provided older adults with a list of specific attributes characterizing SA, and asked them to rate the importance of each attribute. An attribute was deemed important to SA if 75 per cent or more of the participants rated it as important. Of the 20 attributes presented, two different cohort groups identified 13 items as being important to SA that fit into four major themes or dimensions of health: physical, functional, psychological (mental), and social.

Table 10.1 Major themes important to successful aging

Knight & Ricciardelli (2003)	Tate et al. (2003)	Reichstadt et al. (2007)
Health	Health	Health/wellness
Happiness	Satisfaction	Engagement/stimulation
Mental capacity	Keeping active (general)	Security/stability
Adjustment	Keeping active (physical)	Attitude/adaptation
Life satisfaction	Positive attitude	
Physical activity	Family	
Close personal relationships	Independence	
Social activity	Keeping active (mental)	
Sense of purpose	Acceptance	
Withdrawal	Moderation	

While a wide variety of themes and/or elements have been identified as being important to successful aging, how these elements have been used in studies of SA has not always been consistent. Elements used to define SA in some studies have been used as predictors of SA in others (cf. Phelan & Larson, 2002). For example, Roos and Havens (1991) used life satisfaction as a predictor of SA, Fisher (1995) defined it as a precursor of SA, and Caspi and Elder (1986) used life satisfaction as their definition of SA. This inconsistency has presented challenges for researchers and clinicians trying to define the term SA, and apply the research in meaningful ways.

In summary, any definition and/or model of SA needs to be multidimensional in order to address the aging person as a whole. Future studies should strive to be clear in their definitions of independent and dependent variables related to successful aging, and these variables should be justified with reference to theoretical frameworks. While there is much to be gained from the use of the SA framework, it is difficult to make comparisons among studies, given the variety of definitions published.

The most empirically sound definition of SA to date has been Rowe and Kahn's model. In 1987, Rowe and Kahn suggested that, given the inherent heterogeneity among older individuals, it was possible to make the distinction between those aging in a 'usual' fashion, which is experiencing typical age-related declines in function, and those aging 'successfully', those showing minimal decline or loss. The concept of SA allows a focus to be placed on those factors that might explain success in the aging process, and provide appropriate targets for interventions. Rowe and Kahn (1997) defined SA as including three main components: low probability of disease and disease-related disability, high cognitive and physical functional capacity, and active engagement with life. It is the combination of all three components that most fully encapsulates the multidimensional concept of SA. Low probability of disease refers not only to the presence or absence of disease, but also to the absence, presence, and/or severity of risk factors associated with disease. High functional capacity refers to the potential for activity; in other words, not what a person does, but what he or she can do. Lastly, active engagement is primarily concerned with interpersonal relationships and productive activities (i.e., activities that have societal value). This three-pronged approach requires a diverse number of variables be quantified in order to achieve SA. These measures range from social measures (social support, happiness, life satisfaction) to productivity measures (hours spent working in the home or at a job, hours spent volunteering), to health-related (activities of daily living, self-reported health) and cognitive measures (memory, reaction time).

While Rowe and Kahn's model has been widely accepted in the literature, it is not without criticism. Strawbridge et al. (2002) suggested that this definition is limiting in that, in order to be classified as aging successfully, an individual must meet strict criteria in each component of the model, which has resulted in low numbers of individuals (20–33 per cent) being classified as SA in previous studies (Guralnik & Kaplan, 1989; Roos & Havens, 1991; Seeman et al., 1993). However, Rowe and Kahn (1997) assert that their primary interest was in those individuals aging 'well' in spite of the aging process, and this broad based definition 'would encourage older individuals to make lifestyle choices that would maximize their own likelihood of aging well and maintaining a high quality of life in old age' (Kahn, 2002, p. 726). While the self-report measure of SA resulted in a higher percentage of individuals being classified as aging successfully (50.3 per cent), this measure is purely subjective and not based on any estimate of function, performance, or cognition. Although it is encouraging that individuals rate themselves as aging successfully, this does not ensure that these individuals are maintaining a high quality of life.

Following an extensive literature review, Rowe and Kahn (1997) reported three important findings regarding SA: (a) intrinsic factors are not the only determinant of risk in advancing age. Extrinsic environmental factors, including lifestyle, play a role in determining risk; (b) with advancing age, the impact of nongenetic factors increases, while the relative contribution of genetics decreases; and (c) characteristics of usual aging are modifiable. Of particular interest to the current paper are the extrinsic factors, one of which is involvement in physical activity.

According to the 2006 Report Card on seniors in Canada, published by the National Advisory Council on Aging, the personal health habit of 'level of physical activity' in seniors was reported as a weakness, and received a grade of C, indicating that significant improvements are required. Despite all that is known about the positive role that physical activity plays in preventing and minimizing the effects of chronic disease, improving mental health, and enhancing physical health, the majority of Canadian seniors are inactive, a situation that has not improved since 2001. The overall inactivity rate in 2005 was 62 per cent, with 55 per cent of men inactive, and an alarming 67 per cent of women. Unfortunately, this inactivity trend is not unique to seniors. Approximately 56 per cent of adults between the ages of 35 and 64 years are inactive, which is equal to the percentage of inactive adults between the ages of 65–74 years. The rate of inactivity is slightly higher for women aged 75 years and older, where 75 per cent report inactivity. What this highlights is that only about 40 per cent of the adult population under the age of 75 years is physically active in Canada. According to the report card, in many cases, physical activity is negated because it is not incorporated into daily living due to a lack of awareness of the importance it plays in maintaining health in later life (for a different perspective, see Ory et al., 2003), or due to ageist and/or negative stereotypes

164

surrounding involvement (see Horton, Chapter 8). Like many modifiable health risk factors, it is never too late to incorporate physical activity into a daily routine, as significant positive gains can be made in muscle strength, aerobic capacity, and bone density (Kelley et al., 2001; Latham et al., 2003; Lemura et al., 2000). Despite high levels of inactivity, 74 per cent of seniors rate their health as good, very good, or excellent, an increase of four per cent from 2001. The high levels of self-reported health support the data presented by both Strawbridge et al. (2002) and Montross et al. (2006) on self-reported levels of successful aging.

While seniors remain largely inactive, several longitudinal studies have examined the role physical activity plays as a positive predictor of SA within a multi-dimensional framework. While none of these studies has been conducted with MAs, they highlight the positive benefits that even moderate levels of involvement can have on the likelihood of achieving SA. Strawbridge et al. (1996) analyzed data based on the Alameda County Study that began in 1965 and tracked 7,000 adults. Follow-up studies with participants aged 65 years and over focused on two main issues: (a) what factors were expected to be associated with subsequent SA, and (b) what effect did SA have on activities and mental outlook. The definition of SA was based on combined performance on both mobility and performance measures. Subjects were aging successfully if they needed no assistance and had no difficulty on any of the 13 activity/mobility measures (i.e., bathing, walking across a room, dressing), plus little or no difficulty on five physical performance measures (i.e., lifting or carrying weights over 4.54 kg, stopping, pulling a large object). The baseline exercise variable, measured in 1984, was 'often walks for exercise', which was associated with increased odds of being categorized as SA in 1996. Data from the 1990 follow-up reported the proportion of successfully aging 75-year-old men and women engaged in activity relative to those not aging successfully. Ninety-four per cent of men aging successfully reported involvement in exercise or sport, compared to 80 per cent of men not aging successfully. For women, the levels of involvement were 90 per cent and 70 per cent respectively.

Using a similar set of activities of daily living, Leveille et al. (1999) defined SA as living to an advanced old age and having little or no disability prior to death. A respondent was considered 'disabled' if they answered 'yes' to needing help bathing, eating, dressing, transferring from a bed to a chair, using a toilet, or walking across a small room. Using a broader definition of physical activity, level of activity was based on the summed frequency of reported walking, gardening, and vigorous exercise. Overall, two-thirds of men (63 per cent) and women (70 per cent) reporting high levels of physical activity at baseline survived to age 80 years. In contrast, only 34 per cent of men and 47 per cent of women in the low activity group survived to 80 years. Among the most active men, 58 per cent were not disabled prior to death compared with 43 per cent of their least active cohort. An even more dramatic difference existed for women, where there was a twofold difference between the highest and lowest activity levels. To examine this further, Leveille et al. identified a group of participants who were free from disability in activities of daily living and needed no help walking 0.8 km or climbing stairs. This more elite group had half the likelihood of having disability in activities of daily living prior to death when compared to the least active cohort. Interestingly, the intermediate level of physical activity offered no clear advantage over inactivity.

Vaillant and Mukamal (2001) followed both college sophomore and core-city male adolescents for 60 years. Their definition of SA focused on maintaining a high level of well-being in physical, mental, and social functioning. Active college males (i.e., who burned >500 kcalories/week) were nearly four times more likely to be classified as happy–well than sick–sad, or prematurely dead. Newman et al. (2003) examined a larger range of caloric expenditure levels for men and women, ranging from >320 kcal in the lowest quintile of activity to >3520 kcal in the highest quintile. Higher levels of physical activity were associated with increased odds for SA, defined as remaining free of cardiovascular disease, cancer, and chronic obstructive pulmonary disease, with intact physical and cognitive functioning. More recently, Baker et al. (in press), using data from the Canadian Community Health Survey, examined the impact of physical activity on a multidimensional measure of SA based on Rowe and Kahn's model. Low activity was defined as daily energy expenditure < 1.5 METs, moderate activity was energy expenditure of 1.5–3.0 METs, and high activity level corresponded to > 3.0 METs daily. Physical activity was a significant predictor of SA, with physically active and moderately active respondents more likely to be aging successfully compared to the low-activity group.

Menec (2003) examined both the quantity and type of activity as predictors of SA as it related to well-being (life satisfaction and happiness), function (physical [IADLs — instrumental activities of daily living] + cognitive [mental status questionnaire]), and mortality. In this study, she assumed that the more activities one participated in per week, the better it was for SA. Overall, 18 activities were divided into three categories: social (visiting family and friends), solitary (collecting hobbies), and productive (volunteer work, housework). On average participants engaged in an average of eight activities per week. Overall activity level was related to feelings of happiness, improved function in IADLs, ADLs, and reduced cognitive impairment and physical difficulties. Most importantly, activity level was also related to reduced mortality. The odds of dying within six years of the initial interview were reduced for individuals with high activity levels.

With regard to the impact of specific activities, individuals who participated in sports/games reported higher levels of life satisfaction and happiness. Furthermore, social, solitary, and productive activities were related to many components of SA. These data suggest that exercise or sport is not the only mode of activity to have a positive impact on SA. Happiness was related to social groups, all solitary activities (handwork hobbies, music/art/theater, reading/writing), and light housework or gardening. Church-related activities, mass activities (i.e., bingo), music/art/theater, volunteer work, and heavy housework were all related to improved function, and church-related activities and light housework were related to reduced mortality. All of these activities support the themes presented earlier in Table 10.1.

Overall, these studies illustrate the positive role physical activity can play in achieving many dimensions of SA. In particular, the work of Leveille et al. (1999) and Baker et al. (in press) suggest that higher levels of physical activity involvement result in higher levels of SA. However, all of these studies must be interpreted with caution. A limitation inherent in all of these studies is that they did not provide detailed enough information on the type of physical activity and its frequency, intensity, or duration. Thus, it is difficult to compare studies, and it is still unknown exactly how much physical activity is required to produce positive outcomes with respect to SA. Future work should seek to define the parameters of involvement. This information is critical to the development of appropriate community-based programs for older adults.

More importantly, these studies are limited in that they have only examined relatively low levels of physical activity and have not considered the positive contributions that sustained involvement in physical activity can play. Both recreational and competitive MAs typically demonstrate extended periods of sustained involvement, and to date, they have been understudied in this respect. While there is a great deal known about the physiological benefits of Masters athletic participation (see chapters 3 and 4), there is relatively little known about the psychosocial, cognitive, and physical health benefits of Masters participation. Given that this population generally represents some of the healthiest older adults, there is much to be gained by studying them in greater detail.

Shephard et al. (1995) measured the long-term health value of sustained involvement in physical activity. They examined a group of endurance-trained MAs (age range: 41–81 years) over a period of seven years. The favorable health outcomes of the MAs were in large part due to the adoption of a healthier lifestyle. They reported lower incidences of heart disease, hypertension, and diabetes in comparison to rates reported in the general population; and they adopted healthier lifestyles, with 91 per cent reporting that they were very interested in good health, only 2.9 per cent reporting that they currently smoked, 86 per cent always wore a seatbelt, and 88 per cent reported they slept well or very well. Thus, on a very general level, MAs do enjoy physical health benefits from their athletics participation.

On a global level, public health policy should continue to address the outcomes of physical, cognitive, and social health through encouraging older adults to participate in physical activity. Much of the influence to participate either recreationally or competitively in physical activity starts at the community level, where a majority of middle-aged and older adults participate in physical activity programs (e.g., community centers, senior centers, shopping malls, and fitness centers). While very few studies have been conducted examining the impact of the physical and social environment on participation in physical activity by older adults, Booth et al. (2000) found that participants aged 65+ years were more frequently active if they reported the regular participation of friends and family, the availability of paths that were safe for walking, access to local facilities, and high self-efficacy. Litwin (2003) examined the association between the type of social network and physical activity in an older adult sample. Adults who belonged to 'diverse' social networks (married, frequent contact with children, moderate degree of contact with neighbors and friends, occasionally attended religious services, participation at social clubs) reported the highest levels of physical activity (55.8 per cent) compared to those who had a 'family only' network (20.2 per cent) or a 'restricted' social network (19.9 per cent), defined as the least amount of social contact (unmarried, almost no contact with neighbors or friends, lowest rate of religious service attendance). Overall, adults who are the most physically active are the most socially connected.

Drawing these ideas back to Rowe and Kahn's model of SA, the role of the community and neighborhood tie in with the 'active engagement with life' component of the model, where social supports are important. This is the component of the model that has received the least empirical testing, and the one that has a huge potential to influence involvement in physical activity.

Overall, the model of SA is a powerful tool for studying healthy aging. While there are many models available, Rowe and Kahn's three-components model allows physical, cognitive, and social variables to be identified and evaluated. Several studies have demonstrated the positive role that physical activity can play in achieving SA, although it is evident that there is more work to be done. In order for public policy to be influenced by the research, a more systematic

168

evaluation of the influence of physical activity is required. A dose-response relationship needs to be considered to allow more accurate predictions on the benefits of physical activity, and MAs can play a role in addressing this issue. A multilevel approach considering the broader role of the community in promoting physical activity will result in a more comprehensive model and a more thorough understanding of the role of social supports. In conclusion, the MA is critical for understanding the potential that can be achieved in terms of physical, cognitive, and social aging. However, it must start at the grassroots level with simply increasing physical activity participation among seniors. There are gains to be made within each component of the SA model, and physical activity has a vital role to play.

REFERENCES

Baker, J.L. (1958). The unsuccessful aged. *Journal of the American Geriatrics Society*, 7, 570–572.

Baker, J., Meisner, B.A., Logan, A.J., Kungl, A.M., & Weir, P. (in press). Physical activity and successful aging in Canadian older adults. *Journal of Aging and Physical Activity*.

Baltes, P.B., & Baltes, M. (1990). Psychological perspectives on successful aging: The model of selective optimisation with compensation. In P.B. Baltes & M.M. Baltes (Eds.), *Successful Aging: Perspectives from the Behavioural Sciences* (pp. 1–36). Cambridge: Cambridge University Press.

Berkman, L.F., Seeman, T.E., Albert, M., Blazer, D., Kahn, R., Mohs, R., Finch, C., Schneider, E., Cotman, C., & McClearn, G. (1993). High, usual, and impaired functioning in community-dwelling older men and women: Findings from the MacArthur Foundation research network on successful aging. *Journal of Clinical Epidemiology*, 46, 1129–1140.

Booth, M.L., Owen, N., Bauman, A., Clavisi, O., & Leslie, E. (2000). Social-cognitive and perceived environment influences associated with physical activity in older Australians. *Preventive Medicine, 31*, 15–22.

Bowling, A., & Dieppe, P. (2005). What is successful ageing and who should define it? *British Medical Journal*, 331, 1548–1551.

Bowling, A., & Illife, S. (2006). Which model of successful ageing should be used? Baseline findings from a British longitudinal survey of ageing. *Age and Ageing*, 35, 607–614.

Caspi, A., & Elder, G.H. (1986). Life satisfaction in old age: Linking social psychology and history. *Psychology and Aging*, 1, 18–26.

Cooper, L.W., Powell, A.P., & Rasch, J. (2007). Master's swimming: An example of successful aging in competitive sport. *Current Sports Medicine Reports*, 6, 392–396.

Depp, C.A., & Jeste, D.V. (2006). Definitions and predictors of successful aging: A comprehensive review of larger quantitative studies. *American Journal of Geriatric Psychiatry*, 14, 6–20.

Ericsson, K.A., & Charness, N. (1994). Expert performance: Its structure and acquisition. *American Psychologist*, 49, 725–747.

Fisher, B.J. (1995). Successful aging: Life satisfaction and generativity in later life. *International Journal of Aging and Human Development*, 41, 239–250.

Fries, J.F. (1980). Aging, natural death, and the compression of morbidity. *New England Journal of Medicine*, 303, 130–135.

Guralnik, J.M., & Kaplan, G.A. (1989). Predictors of healthy aging: Prospective evidence from the Almeida County Study. *American Journal of Public Health*, 79, 703–708.

Hawkins, S.A, Wiswell, R.A., & Marcell, T.J. (2003). Exercise and the master athlete — a model of successful aging. *The Journals of Gerontology*, 58A, M1009–1011.

Health Canada (2002). *Canada's Aging Population* (Cat. H39–608/2002E). Ottawa, Ontario: Health Canada.

Kahn, R.L. (2002). On successful aging and well-being: Self-rated compared with Rowe and Kahn. *The Gerontologist*, 42, 725–726.

Kelley, G.A., Kelley, K.S., & Tran, Z.V. (2001). Resistance training and bone mineral density in women: a meta-analysis of controlled trials. *American Journal of Physical Medicine and Rehabilitation*, 80, 65–77.

Knight, T., & Ricciardelli, L.A. (2003). Successful aging: Perceptions of adults aged between 70 and 101 years. *International Journal of Aging and Human Development*, 56, 223–245.

Latham, N., Anderson, C.S., Bennett, D., & Stretton, C. (2003). Progressive resistance strength training for physical disability in older people. *Cochrane Database of Systematic Reviews*, Issue 2, CD002759.

Lemura, L.M., von Duvillard, S.P., & Mookerjee, S. (2000). The effects of physical training of functional capacity in adults: Ages 46 to 90: A meta-analysis. *Journal of Sports Medicine and Physical Fitness*, 40, 1–10.

Leveille, S.G., Guralnik, J.M, Ferrucci, L., & Langlois, J.A. (1999). Aging successfully until death in old age: Opportunities for increasing active life expectancy. *American Journal of Epidemiology*, 149, 654–664.

Litwin, H. (2003). Social predictors of physical activity in later life: The contribution of social-network type. *Journal of Aging and Physical Activity*, 11, 389–406.

Menec, V.H. (2003). The relation between everyday activities and successful aging: A 6-year longitudinal study. *The Journals of Gerontology*, 58B, S74–S82.

Montross, L.P., Depp, C., Daly, J., Reichstadt, J., Golshan, S., Moore, D., Sitzer, D., & Jeste, D.V. (2006). Correlates of self-rated successful aging among community dwelling older adults. *American Journal of Geriatric Psychiatry*, 14, 43–51.

National Advisory Council on Aging. *Seniors in Canada: 2006 Report Card* (Cat.HP30–1/2006E). Ottawa, Ontario: National Advisory Council on Aging.

Newman, A.B., Arnold, A.M., Naydeck, B.L., Fried, L.P., Burke, G.L., Enright, P., Gottdiener, J., Hirsch, C., O'Leary, D., & Tracy, R. (2003). 'Successful aging': Effect of subclinical cardiovascular disease. *Archives of Internal Medicine*, 163, 2315–2322.

Ory, M., Hoffman, M.K., Hawkins, M., Sanner, B., & Mockenhaupt, R. (2003). Challenging aging stereotypes: Strategies for creating a more active society. *American Journal of Preventive Medicine*, 25(3), 164–171.

Palmore, E. (1979). Predictors of successful aging. *Gerontologist*, 19, 427–431.

Peel, N., Bartlett, H., & McClure, R. (2004). Healthy ageing: How is it defined and measured? *Australian Journal on Ageing*, 27(3), 115–119.

Pennix, B.W.J.H., van Tilburg, T., Kriegsman, D.M.W., Deeg, D.J.H., Boeke, A.J.P., & van Eijk, J.T.M. (1997). Effects of social support and personal coping resources on mortality in older age: The Longitudinal Aging Study, Amsterdam. *American Journal of Epidemiology*, 146, 510–519.

Phelan, E.A., & Larson, E.B. (2002). 'Successful Aging' — Where next? *Journal of the American Geriatrics Society*, 50(7), 1306–1308.

Phelan, E.A, Anderson, L.A., LaCroix, A.Z., & Larson, E.B. (2004). Older adults' views of 'successful aging' — How do they compare with researchers' definitions? *Journal of the American Geriatrics Society*, 52, 211–216.

Pressey, S.L., & Simcoe, E. (1950). Case study comparisons of successful and problem old people. *Journal of Gerontology*, 5, 168–175.

Reed, D.M., Foley, D.J., White, L.R., Heimovitz, H., Burchfiel, C.M., & Masaki, K. (1998). Predictors of healthy aging in men with high life expectancies. *American Journal of Public Health*, 88(10), 1463–1468.

Reichstadt, J., Depp, C.A., Palinkas, L.A., Folsom, D.P., & Jeste, D.V. (2007). Building blocks of successful aging: A focus group study of older adults' perceived contributors to successful aging. *American Journal of Geriatric Psychiatry*, 15, 194–201.

Roos, N.P., & Havens, B. (1991). Predictors of successful aging: A twelve-year study of Manitoba elderly. *American Journal of Public Health*, 81, 63–68.

Rowe, J.W., & Kahn, R.L. (1987). Human aging: Usual and successful. *Science*, 237 (4811), 143–149.

Rowe, J.W., & Kahn, R.L. (1997). Successful aging. *The Gerontologist*, 37, 433–440.

Salthouse, T.A. (1991). Cognitive facets of aging well. *Generations*, 15, 35–38.

Schulz, R., & Curnow, C. (1988). Peak performance and age among superathletes: Track and field, swimming, baseball, tennis, and golf. *Journal of Gerontology: Psychological Sciences*, 43, P113–P120.

Schulz, R., & Heckhausen, J. (1996). A life span model of successful aging. *American Psychologist*, 51, 702–714.

Seeman, T.E., Rodin, J., & Albert, M. (1993). Self-efficacy and cognitive performance in high-functioning older individuals: MacArthur studies of successful aging. *Journal of Aging and Health*, 5, 455–474.

Seeman, T.E., Berkman, L.F., Charpentier, P.A., Blazer, D.G., Albert, M.S., & Tinetti, M.E. (1995). Behavioral and psychosocial predictors of physical performance: MacArthur studies of successful aging. *Journal of Gerontology: Medical Sciences*, 50A, M177–183.

Shephard, R.J., Kavanagh, T., Mertens, D.J., Qureshi, S., & Clark, M. (1995). Personal health benefits of Masters athletic competition. *British Journal of Sports Medicine*, 29(1), 35–40.

Simonton, D.K. (1988). Age and outstanding achievement: What do we know after a century of research? *Psychological Bulletin*, 104, 251–267.

Strawbridge, W.J., Cohen, R.D., Shema, S.J., & Kaplan, G.K. (1996). Successful aging: Predictors and associated activities. *American Journal of Epidemiology*, 144, 135–141.

Strawbridge, W.J., Wallhagen, M.I., & Cohen, R.D. (2002). Successful aging and well-being: Self-rated compared with Rowe and Kahn. *The Gerontologist*, 42, 727–733.

Tate, R.B., Leedine, L., & Cuddy, T.E. (2003). Definition of successful aging by elderly Canadian males: The Manitoba follow-up study. *The Gerontologist*, 43, 735–744.

Vaillant, G.E., & Mukamal, K. (2001). Successful aging. *The American Journal of Psychiatry*, 158, 839–847.

World Health Organization (1948). *Constitution of the World Health Organization*. Geneva: World Health Organization.

CHAPTER ELEVEN

INJURY EPIDEMIOLOGY, HEALTH, AND PERFORMANCE IN MASTERS ATHLETES

WILLIAM J. MONTELPARE

At the simplest level, epidemiology is an investigative process in which the distribution and determinants of health-related events are studied (MacMahon & Pugh, 1970). Within the population of Masters Athletes, epidemiological methods provide valuable information about specific health outcomes. Most important, the applications of epidemiological methods help researchers understand potential health risk factors as well as likely protective factors for positive health status (Young, 1998). Epidemiological methods can help researchers determine the temporality of events that lead to measurable health outcomes. Although injury epidemiology holds great promise for increasing our understanding of injuries and other factors affecting performance decline in older athletes, this type of research is underutilized.

INJURY EPIDEMIOLOGY

Typically in studies of exercise epidemiology, the approximation of risk is given as a rate estimate measured as either incidence (a time dependent rate) or prevalence (a summative rate). These rate estimates are largely drawn from surveillance systems designed for contributions by athletic trainers or sport/team managers. Such systems of injury surveillance are specifically intended to monitor events that result from either training regimens or competitive performances, and in so doing, record morbidity (illness/injury) and, in rare circumstances, mortality of the athletes. For the purposes of our discussion of the Masters Athlete population, most often the epidemiological measure of interest is a reported injury. When a sport-related injury is reported, a rare event in itself at any level of competition, the injury is typically due to overuse and may be directly attributable to the individual's training regimen or performance schedule (see Fell & Williams, Chapter 6).

Epidemiological applications to the study of sports injury generally include an examination of the distribution of injuries and the investigation of injury determinants, with a general view toward injury prevention. Epidemiological investigations of determinants for injury among Masters Athletes include but are not limited to: sport-specific training for a prolonged period (years versus months), and exposure to various environmental conditions (heat, cold, altitude, and air quality). Yet an important consideration in evaluating predisposing characteristics among Masters Athletes are the intrinsic characteristics and training behaviors of the Masters Athletes during the process of healthy aging. The latter characteristic, a departure from the processes of normal aging, may be most important as it may help to explain the variability of individuals within the Masters Athlete cohort in comparison to similar-aged individuals from a normal aging population.

The study of injury epidemiology and the underlying theories of injury causation have undergone several transformations. Initially, injuries were ascribed to accidents, and accidents were generally accepted as a function of implicit behaviors by an individual. In these early descriptive models of injury prevalence (Gibson, 1961; Haddon, 1968), there was a notion that causal mechanisms were random events which led to the observed outcomes. Similarly, given that injury-causing mechanisms occurred at random, there was a concomitant acknowledgment that injuries resulted from an accepted level of unexplained risk. Subsequent work using the models proposed by Gibson (1961) and Haddon (1968) removed this veil of mystique and advanced the application of the *epidemiological triad* to explain injury dynamics.

The epidemiological triad consists of host, agent, and environment (Andersson & Menckel, 1995; see Figure 11.1). In this triad the term *host* refers to the individual. The Masters Athlete is the host, a convention that is similar and widely

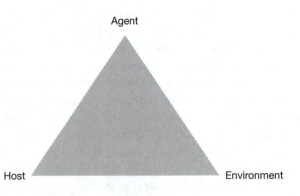

Figure 11.1 The epidemiological triad

accepted within infectious disease models (Young, 1998). Several variables can be evaluated in relation to the characteristics of the host. For example, in describing the cohort of Masters Athletes, the host can be separated by sex (males versus females), sport classifications (team games versus individual pursuits), and of course classification across specific age groups (often in quintiles, such as 35–39, 40–44, 45–49). Within each of these classifications, there can be further subdivisions, organizing sport by the specific type of metabolic demand (continuous versus intermittent) or based on oxidative metabolic demands (aerobic versus anaerobic) or energy system utilization in relation to neural–muscular recruitment (fast glycolytic, fast-oxidative glycolytic, or slow twitch-endurance). Given the size of the athlete sample and the sensitivity of measurements (e.g., identifying individuals that truly meet the selection criteria and specificity of measurements, or removing individuals that truly do not meet the selection criteria of the classification strategies), subdivisions of the Masters Athlete cohort can yield valuable descriptions of the characteristics of the host.

The term *agent* within the epidemiology triad refers to the causal stimulus, and can be considered as the impetus that leads to a reported injury. Specifically in relation to injury epidemiology and the Masters Athlete, one could consider that an injury occurs when the energy transferred from a given source to the host exceeds the capacity of the tissue receiving the energy to adapt or accommodate the magnitude of the stimulus. Under this definition, a mechanism of transmission is not only needed to transfer the energy from a source to the host within a given environment, but such a mechanism must be observable.

With regard to Masters Athletes, the mechanism of injury can be overt, as in a collision between two individuals, an athlete falling off their bicycle, or an athlete being struck by a projectile. However, the mechanism can also be undistinguishable and therefore unattributed, such as the effects of overtraining, or as the result of participating in competition, which may lead to muscle damage at the cellular level (conditions which are described in detail by Fell and Williams, Chapter 6).

Finally, the term *environment* within the epidemiology triad refers to the surroundings, the situation, or the setting. The environment can refer to the training environment which describes the characteristics of the landscape (topography), the equipment, or the air or water quality. Likewise, the environment can describe the social milieu that defines the atmosphere or ambience of the training program. Determining whether the environment is competitive or collaborative provides important information about factors that may help to explain injuries that occur during training.

EPIDEMIOLOGICAL MODELS

The work of Gibson (1961) and Haddon (1968) emphasized the importance of recognizing the role of energy transfer in injury outcomes, as well as the need to consider the occurrence of other actions at the time the injury occurred. Despite this work, there continue to be sport enthusiasts who readily accept the prevalence of injuries as *part of the game — an expectation related to participation,* and therefore without a need for explanation or a need to consider prevention strategies to control the outcomes. Indeed, the belief that injuries are 'natural' or inevitable may be a predisposing characteristic that elevates injury likelihood.

Hagberg et al. (1997) extended the understanding and recognition of several predisposing factors advanced by Gibson (1961) and later explained by Haddon's injury matrix (Haddon, 1968) in a conceptual model specifically relating to occupational settings. As Haddon (1968) demonstrated using a relational matrix, it is possible to describe injuries as a function of interactions between the level of exposure to energy transfer from both a frequency and magnitude perspective, especially in relation to host and agent interaction. The application of the Haddon Matrix to understanding injuries among Masters Athletes serves as an effective approach by which to describe events that predispose the Masters Athlete to injury. Likewise, this approach helps to explain an individual's propensity to report injuries as well as the way an individual might classify an event as being an injury.

A central tenet to understanding the occurrence of injury versus the more likely reported cumulative trauma outcomes (i.e., overuse conditions) is that the events underlying the injury event are probabilistic (based on a chance that they will transpire), and therefore the occurrence of the injury is a result of the interaction of several variables at the point of the incident (Hagberg et al., 1997). In Hagberg's conceptual model, where injury outcome is shown to result from the interaction of exposure, stimulus energy and behaviors/conditions, individuals will have a low frequency of occurrence of exposure to high energy transfer, such as being struck by an automobile while training (see Figure 11.2). This outcome may be described as a rare event resulting from the interaction between energy transfer, host behavior, and environmental conditions. Similarly, the Masters Athlete is not as likely to report events that occur at a high frequency but with low levels of energy transfer. For example, injury situations that do not result in loss of exercise/training time or event participation (e.g., minor collisions with opponents, abrasions from bicycle falls, or feelings of strain that occur while participating in extended exercise/physical activity).

william j. montelpare

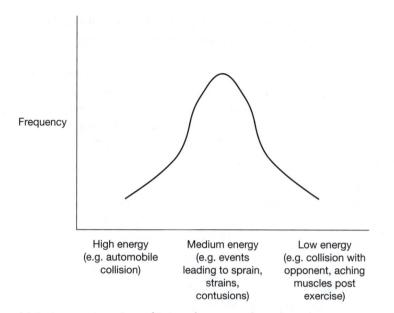

Figure 11.2 A representation of injury frequency based on the suggestion of the Hagberg Model

Conversely, in the Hagberg model it is more likely that, within a Masters Athlete cohort, there is a frequent occurrence of low-energy transfers that can accumulate over time and go undetected or ignored by the individual. These types of injury reports often lead to the description of a cumulative type disorder, although Hagberg explained that this reference to sub-clinical injury needs clarification with the definition as well as with the description of the injury terminology. Consistent with this injury dynamic is the notion of an energy transfer exposure period, which is referred to by Hagberg as the latency period; that is, the time between the first exposure to an injury agent and the manifestation of the injury outcomes by the host (the individual).

Consider, for example, muscle tissue damage that may be a result of strain or sprain during training. Fell and Williams (Chapter 6) describe the concept of such injury outcomes in physiological terms and conclude that confounding variables, such as age-related declines in training loads, must be controlled in order to determine the extent to which recovery from muscle damage caused by training can be truly measured. The inability to fully recover from training-induced muscle damage can not only prolong the recuperation period, but, if left untreated, or if the athlete re-establishes their training regimen too soon after the injury, the outcome can lead to further damage that may cause a cessation of physical activity.

The time between first exposure to an injury stimulus and the bona fide signs of an injury may play an important role in quantifying the actual duration between exposure to an agent and the likelihood of an injury occurring. This information would be valuable for the detection and prevention of injuries resulting from training or competition-related stressors. In cases of concussion, for example, recognizing that a lag exists between first impact and a subsequent sign of traumatic brain injury is essential to the prevention of second impact syndrome, an event with consequences ranging from minimal reductions in neuropsychological performance to death.

Finally, while the model developed by Hagberg et al. is comprehensive and efficient in describing injuries related to occupational settings, and there is transferability of the central concepts to the epidemiological study of injuries among Masters Athletes, one must recognize the multifactor nature of injuries and the possibility that the various determinants of an injury outcome within the Masters Athlete cohort are not captured by this particular model. Continued research into levels of exposure, mechanisms of injuries, and predisposing characteristics of the host and the environment to the injury agent are among recommendations for continued research in this area.

ASSOCIATION AND CAUSATION

Association refers to the measurable relationship, or lack thereof, between two or more variables. Causation is defined as a logical series of events beginning with a condition (stimulus) or characteristic that, when introduced to an individual or cohort (sample, population), produces a measurable outcome. The terms association and causation are interdependent in epidemiological study. The suggestion that a stimulus is causal depends on the extent to which there is an identifiable relationship or association between an individual's exposure to the suspected stimulus and the measurable incidence or prevalence of the outcome/injury. In keeping with the epidemiology triad as a fundamental model of injury prevalence among Masters Athletes, measurements of association require that one of the variables will be an injury-causing stimulus (an agent), while another will be the observable injury outcome. Similarly, when assessing the causal impact of a noted stimulus, the generally accepted approach is to consider the stimulus as a sufficient cause when the presence of the stimulus leads to a measure of the outcome.

Applications of data mining approaches to identifying patterns within a set of probable injury determinants is an essential step in advancing our understanding of the epidemiology of injuries among the Masters Athlete cohort. Through the

178

processes of data mining, researchers can target specific activities and identify behaviors of individuals in these activities so that they can establish patterns of association and measures of causation with estimable probabilities. Using large-scale research data sets, injury registries, and/or injury surveillance databases, researchers can work through data mining exercises to create data models in which several variables can be evaluated for the relative contribution of each as a predictor to an observed injury outcome. As a rule, epidemiologists do not consider any single study as sufficient evidence to support a causal inference, especially when considering observational research often conducted using an epidemiological approach. Rather, it is generally recognized that, while there may be reasons to explain a relationship, it is equally likely that there are reasons to explain the lack of a relationship.

RISK FACTORS VERSUS PROTECTIVE FACTORS

When stimuli are continuously associated with observed outcomes/injuries, then these stimuli are referred to as risk factors. Risk factors describe the characteristics which underlie the probability or likelihood of an outcome/injury. Two essential rate estimates that help to describe risk factors in epidemiological studies are morbidity and mortality. Estimates of morbidity and mortality are expressed as outcome rates in relation to the number of individuals at risk. Generally, these rates are standardized across a number of specific cohorts and are expressed as proportional rate estimates that describe the number of individuals demonstrating either the morbidity or mortality outcome in relation to all those individuals that could demonstrate the outcome. An example of a mortality rate is the age-standardized mortality rate, where the number of deaths is reported for a specific age group within a given cohort of the population. An example of how this data is useful was shown by Sandvik et al. (1993) in comparing the differences in relative risk of death across different fitness levels. In their study, Sandvik et al. showed that individuals in the highest fitness cohort were half as likely to die from cardiovascular-related diseases than individuals in the next lowest cohort, even after adjusting the data for smoking status and age.

Estimates of morbidity are typically outcome specific and can be used to describe health status for a cohort in relation to a given disease/illness/injury. Morbidity rates are presented as either incidence estimates — the number of new cases observed in relation to all those individuals at risk within a given time period; or prevalence estimates — the number of cases observed in relation to all those individuals at risk. Injury rates reported for the Masters Athlete cohort represent a prevalence morbidity rate, while injury rates reported for the Masters Athletes competing at the Seniors Olympics represent an incidence morbidity rate.

EPIDEMIOLOGY AND THE STUDY OF MASTERS ATHLETES

Masters Athletes represent a special cohort of age-group athletes. In general, the potential that the study of injury epidemiology among the Masters Athlete cohort holds has been largely untapped by researchers. Masters Athletes should be considered as outliers to the normal aging distribution, and therefore outliers to the normal distribution of age-group athletes. As Tanaka and Seals (2008) suggest, the elite Masters Athlete is an individual who demonstrates the characteristics of exceptional aging (also see Weir, Chapter 10). Understanding the rates and proportions of injuries within the Masters Athlete population, and developing a sound understanding of those factors that may lead to reports of injury within this population, are necessary to understanding factors that may not only inhibit but also end a Master Athlete's involvement in exercise training and competition.

If we consider that epidemiological study of the Masters Athlete will be limited to reporting on the outcome/injury using the tools of epidemiological categorization and description (e.g., morbidity, mortality, incidence, and prevalence), then a major epidemiological question is whether or not the outcome/injury is a result of: (a) the training and performance as a Masters Athlete, (b) training and performance prior to reaching the age of a Masters Athlete, or (c) some other factor that may or may not be related to the aging process.

Large-scale injury epidemiology studies of Masters Athletes are rare. Most data collection for injury studies of this population has been through participant surveys because there is no consistent injury-reporting system across the various Masters Athlete sports. One such survey was conducted on some 2,886 runners by McKean et al. (2006). In their study, individuals described, retrospectively, injuries which resulted from their running training or performance experience. In a comparison of respondents by age group, the authors found that the prevalence of injuries was higher among the Masters runners. The cohort of respondents aged 40 years and older reported not only more injuries but more multiple injuries than runners who were less than 40 years of age. Likewise, McKean et al. stated that injuries identified by the Masters Athletes were most often soft-tissue injuries of the calf, hamstring, or Achilles tendon. The authors concluded that, regardless of injury type or injury mechanism, the risk of injury increased in relation to volume of training, and, more specifically, the risk of injury increased in proportion to the volume of running frequency

william j. montelpare

per week. Among Masters Athletes, the specific mechanisms leading to overuse injuries are varied; however, in general, one can consider that the injury stimulus is too much repetitive movement on tissues that are incapable of handling the specific type of physical stress (DiPietro, 2007).

Heading a soccer ball is an event which is often practiced by Masters-level soccer players. This specific activity describes a sequela of events among the Masters Athlete cohort that differs from those events that may be observed in a younger athlete cohort performing the same activity. Consider the application of the injury epidemiological triad. In this scenario, the athlete is the host, the action of heading the ball is the agent, and the game of soccer is the environment. In a comparison study of soccer players and swimmers, Downs and Abwender (2002) compared neuropsychological performance measures for a group of college-age (average age 19 years) and a group of older soccer players (average age 41 years) to similar age-matched groups of swimmers (average ages 19 and 42, respectively). The authors found that older soccer players scored lower on neuropsychological performance measures than both younger soccer players and age-matched swimmers. While sufficient evidence does not currently exist to establish that heading a ball alone is an agent that can lead to mild traumatic brain injury (MTBI), the authors suggested that a dose-response relationship may exist between heading a ball and MTBI prevalence. That is, as a result of continued heading of a soccer ball, aging players — and especially those individuals with a history of concussion — could be at a higher risk for MTBI and long-term cognitive impairment than younger participants or individuals who do not participate in this type of activity.

Korpelainen et al. (2001) used the data mining approach of variable extraction to describe the association between clinically diagnosed stress fractures of the lower leg among runners as the injury outcome variable with a list of possible injury predictor variables: nutrient intake, training history, balance while standing, foot structure, pronation and supination of the ankle, dorsi-flexion of the ankle, leg-length inequality, and range of hip rotation. Korpelainen et al. used a sample size of 34 athletes drawn from a sports injury clinic database containing some 12,150 patients treated over a 13-year period. The average age of the selected patients was 20 years at the point of first injury report, and participants in the treatment group were matched to a smaller cohort of control athletes on the basis of age, sex, body mass index, and sport type.

All participants completed a questionnaire to provide information about nonphysical measures so that combining these data with the objective clinical measures enabled the researchers to run both parametric and non-parametric statistical analyses. The data were derived from an array of variables of mixed types. Some of the variables collected were measures specific to biomechanical functioning, while others were related to general lifestyle, training regimen, and medical history. The authors concluded that, in addition to reports of physical problems such as stress fractures or menstrual irregularities among female participants, there was a measurable relationship between volume of training mileage and risk of recurrent injuries.

The statistical approaches and level of enquiry practiced by Korpelainen et al., although somewhat cursory by data mining standards, reflect the potential for more involved analyses of measurements collected from individuals using large-scale data sets. Given that there were several variables collected from the larger population of patients in the injury clinic database, data mining approaches provide researchers with the capability to combine measurements across different cohorts and illustrate the similarities and differences in the dynamics that lead to injury. Since some epidemiological investigations of association suggest that any seemingly apparent relationship between an exposure/agent and an injury/outcome could have an alternative reason/explanation, there is a need to use more advanced data mining approaches to the study of large-scale surveillance data sets. Merely providing simple correlation estimates between agent and outcome may overlook hidden influencing factors. Data mining applications enable the researcher to evaluate the appropriateness of various alternative explanations, and adjust for chance, bias, systematic error, or confounding. By increasing our understanding of injuries through the rigorous investigation of injury models for association, we will ultimately enhance our ability to understand the mechanisms and predisposing factors to an injury outcome.

DIRECTIONS FOR FUTURE RESEARCH

The other chapters in this book have clearly established that Masters Athletes are a distinct cohort of individuals with physical activity behaviors that are recognizably different from the typical aging adult. Whereas the typical aging adult rarely considers physical activity as a daily recreational pastime, the Masters Athlete is focused, determined, habitual, and committed to their training regimen.

As Larson and Bruce (1987) suggested, measurable physiological adaptations can result from regular compliance to sub-maximal work. However, given that the Masters Athlete's training regimen often exceeds sub-maximal levels and extends beyond recommended minimal participation levels (ACSM, 1998), there is an increased likelihood that the risk for morbidity and, in rarely reported cases, mortality, may be greater than a comparison cohort of individuals similar in age but physically active at a lower frequency, intensity, and duration.

Without question, the behavior of the Masters Athlete in regard to training regimens, motivation to participate, and competition warrants further study. While there is a need to continue our ongoing evaluation of physiological adaptations of the aging individual's response to training, as well as the need to develop an understanding of an individual's motivation to compete, there is also a need to determine the extent to which these behavioral characteristics influence the epidemiological measures of morbidity and mortality.

The first step in the process of understanding the epidemiology of the Masters Athlete is to establish standardized methods of outcome surveillance. The approach to comprehensive injury surveillance should be based on registration systems which identify individuals as Masters Athletes (beyond merely recreational enthusiasts) who maintain regular training schedules and who participate in recognized competitions. Standardized surveillance systems can provide stable and valid data of morbidity and mortality, but only if they are populated with data on a regular basis by qualified data-entry personnel. Surveillance systems that store incomplete records, suspect data, and cases that are lost to follow-up provide no information for research and programming purposes.

Conversely, standardized surveillance systems can provide accurate participant data that are used to establish the true number of individuals participating, therefore providing an accurate estimate of risk ratios. While a standardized surveillance system can expedite the recording of outcomes attributed to training and competition, these standardized systems for injury data collection will require that the records are continuously updated and validated, or they will become obsolete very quickly.

Currently there is no single repository that best describes the epidemiology of Masters Athletes because there is no consistent, comprehensive, and inclusive system of epidemiological outcome reporting. Injuries can and do occur within the cohort of Masters Athletes, but all too often injuries are not reported to any system of surveillance. This lack of a standardized injury reporting system inhibits the development of knowledge about the critical issues which underlie cause, treatment, and return to participation. A central repository for injury reporting and tracking could provide essential follow-up information about the individual with respect to epidemiological characteristics. Further, secondary statistical analyses of information within an injury repository could help to create programs that would reduce recognized risk factors for injuries among Masters Athletes. In particular, the data provided by analyses of the injury reporting system could help to create a standardized definition of an injury that is event specific and is based on criteria that are easily understood to define and describe injuries.

A graphical profile of physical activity involvement (scaled on the horizontal axis) and epidemiological outcome (scaled on the vertical axis) most often reflects a line graph that is represented as a j-curve (see Figure 11.3). That is, low physical activity (left side of the horizontal axis) correlates with a higher prevalence of morbidity/mortality (the vertical axis), while increased physical activity correlates to a decline in morbidity/mortality. At a specific point on the horizontal axis representing increasing volumes of physical activity/training, there begins a steady increase in the outcome measures (morbidity and mortality) as a direct relationship to increased training volumes.

A standardized surveillance system can provide essential evidence to determine the slope of the outcome response in regard to training regimens. Moreover, given the large populations of athletes competing in organized international events, such as the World Masters Games and Senior Olympics, as well as national championships, a standardized injury reporting system may be possible provided that organizational structures and sport governing bodies emphasize its importance both philosophically and through increased resources. As the reservoir of Masters Athlete injury data increases, researchers will be able to create specific statistical models of injury outcomes, which may provide essential probabilistic evidence to inform knowledge of injury risk factors in Masters Athlete cohorts. Further, the evidence gained from statistical models of injury dynamics in Masters Athletes

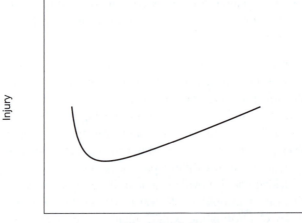

Physical activity volume

Figure 11.3 Relationship between physical activity involvement and injury outcome

184

will enable coaches and administrative bodies to create injury prevention strategies and programs that not only reduce the risk of injury within the Masters Athlete population but also reduce the burden of injury on society.

REFERENCES

ACSM (1998). Exercise and physical activity for older adults. *Medicine and Science in Sports and Exercise*, 30, 992–1008.

Andersson, R. & Menckel, E. (1995). On the prevention of accidents and injuries. A comparative analysis of conceptual frameworks. *Accident: Analysis and Prevention*, 27l, 757–768.

DiPietro, L. (2007). Physical activity, fitness and aging. In C. Bouchard, S. Blair, & W.L. Haskell (Eds.), *Physical Activity and Health* (pp. 271–285), Windsor: Human Kinetics.

Downs, D.S., & Abwender, D. (2002). Neuropsychological impairment in soccer athletes. *Journal of Sports Medicine and Physical Fitness*, 42, 103.

Gibson, J. (1961). The contribution of experimental psychology to the formulation of the problem of safety: A brief for basic research. In L.W. Mayo (Ed.), *Behavioral approaches to accident research* (pp. 77–89). New York: Association for the Aid of Crippled Children.

Haddon, W. (1968): Advances in the epidemiology of injuries as a basis for public policy. *Public Health Reports*, 95, 411–421.

Hagberg, M., Christiani, D., Courtney, T.K., Halperin, W., Leamon, T., & Smith, T.J. (1997). Conceptual and definitional issues in occupational injury epidemiology. *American Journal of Industrial Medicine*, 32, 106–115.

Korpelainen, R., Orava, S., Karpakka, J., Siira, P., & Hulkko, A. (2001). Risk factors for recurrent stress fractures in athletes. *American Journal of Sports Medicine*, 29, 304–310.

Larson, E.B., & Bruce, R.A. (1987). Health benefits of exercise as an adult. *Archives of Internal Medicine*, 147, 353–356.

MacMahon, B., & Pugh, T. (1970). *Epidemiology: Principles and methods*. Boston: Little & Brown.

McKean, K.A., Manson, N.A., & Stanish, W.D. (2006). Musculoskeletal injury in the Masters runner. *Clinical Journal of Sport Medicine*, 16,149–154.

Sandvik, L., Erikssen, J., Thaulow, E., Erikssen, G., Mundal, R., & Rodahl, K. (1993). Physical fitness as a predictor of mortality among healthy, middle-aged Norwegian men. *New England Journal of Medicine*, 328(8), 533–537.

Tanaka, H., & Seals, D.R. (2008). Endurance exercise performance in Masters athletes: Age-associated changes and underlying physiological mechanisms. *Journal of Physiology*, 586, 55–63.

Young, T.K. (1998). *Population health: Concepts and methods*. New York: Oxford University Press.

CHAPTER TWELVE

THE FUTURE OF MASTERS GAMES
Implications for policy and research

ROY J. SHEPHARD

The World Masters Games (WMG) saw their debut in Toronto in 1985, with 8,305 participants from 61 countries participating in 22 types of sport (Kavanagh et al., 1988a; Kavanagh et al., 1988b; Shephard et al., 1995). Other cities have subsequently hosted the WMG at three- to four-year intervals, with attendance apparently depending in part on the choice of venue (see Chapter 1). The most popular meets to date have been in Brisbane (1994: 24,500 participants from 74 countries competing in 30 disciplines) and Melbourne (2002: 24,886 participants from 98 countries competing in 26 disciplines). Sydney (2009) hopes for a similar interest, with some 30,000 from 100 countries competing in 15 core and 13 optional disciplines. These numbers are quite large in comparison with the Beijing Olympics of 2008 (10,500 athletes participating in 28 sports) and the Beijing Paralympics (4,200 participants in 20 sports). One of the main differences between these events and the WMG is a somewhat smaller international involvement in the WMG (around 100 nations, compared with the 205 nations involved in the Olympics and 148 nations in the Paralympics of 2008). Nevertheless, the WMG is a well-established phenomenon, and its future poses some significant problems for the academic community in terms of both social policy and research priorities.

DIRECT POLICY IMPLICATIONS

Although the WMG attracts a large number of participants, the spectators are drawn largely from friends, relatives and fellow competitors. The meets have attracted relatively little interest from either government or media. There is a stark contrast between the estimated $42 billion spent in staging the Beijing Olympics and the recent grant of $100,000 that the New South Wales government thought appropriate to offer to the organizers of the Sydney WMG. Possible reasons for the low level of political and popular interest include the facts that events are open to applicants at all levels of ability, and the nature of some of the events, such as lawn bowling, hardly makes for dynamic and exciting television. Further, the number of age categories is such that the field of contestants in any one event may be quite small, particularly in the older age groupings. As yet, it remains unclear whether the number of entrants in older age categories will be boosted as older enthusiasts age. The similarity of numbers between Brisbane (1994) and Melbourne (2002) argues against such a hypothesis, and in the absence of more generous governmental sponsorship, the rising cost of air fares and hotel accommodation may tend to restrict future meets. Certainly, in most nations, the number of people actively involved remains too small to have any important *direct* impact on population health.

How far do the exploits of older participants affect public health indirectly (as proposed by Horton, Chapter 8)? Anecdotal reports suggest that individuals such as Roland Michener (Governor General of Canada, 1967–1974) and Jack Rabbit Johansson (still actively cross-country skiing on his hundredth birthday) have attracted some popular interest. However, whether the majority of seniors are aware of and inspired by such individuals is not known. How far is the general public aware of Masters sport? And how has the existence of this type of involvement altered personal health behavior, if at all? Has Masters sport really had a positive effect on stereotypes of aging? Where do seniors look for role models (Lockwood et al., 2005)? Do seniors see Masters competition as something that they can achieve? Or are they intimidated by such feats (Ory et al., 2003)? Can the example of Masters Athletes change overall attitudes towards seniors, as has been suggested (Levy & Banaji, 2002)? At present, the evidence tends to be sketchy and anecdotal, with much of the available information on role modeling obtained from much younger age groups. But if further enquiry does reveal positive effects from Masters competition, at least two further questions must be answered. Does Masters sport promote the most effective pattern of physical activity for enhancing the health of middle-aged and elderly individuals? And if Masters sport is to be advocated as a method of exercise promotion, how do the costs of Masters sports (in terms of training, travel to training centers and meets, and a need for special facilities) compare with the overall costs of alternative forms of advocacy?

EFFECTS ON OVERALL HEALTH AND THE AGING PROCESS

Does training for and participation in the WMG enhance overall health? Typically, studies of physical activity and health have focused on overall life expectancy

(Paffenbarger et al., 1994), although particularly from the standpoint of the elderly, the quality-adjusted life expectancy (QALE) is more important than the mere years of survival (Shephard, 1996).

Overall life expectancy

Longitudinal studies of some categories of athletes have shown a pattern of participation until middle-age; subsequently, training may cease without a corresponding modification of food intake, and at this stage the former athlete becomes fatter than sedentary peers, with a corresponding deterioration in life expectancy (Montoye et al., 1957). In contrast, some studies of endurance athletes from Finland have found a several-year advantage of life expectancy relative to the general population (Karvonen et al., 1974; Sarna & Kaprio, 1994). It has been argued that the reason for this advantage has been a focus on sports such as cross-country skiing that the athletes would be likely to continue into old age. Investigators have recognized some of the potential confounding variables in endurance athletes, such as an ectomorphic body build and lifelong abstinence from cigarette smoking, but analyses to date have not succeeded in clarifying how far any benefits are due to continued sport participation. There remains a need for careful longitudinal studies on samples matched for smoking habits and body build, comparing mortality at selected ages between various categories of WMG participant, those who engage habitually in other forms of physical activity such as brisk walking, and sedentary individuals. If a survival advantage can be shown for some or all WMG categories, the question remains whether a similar or even a larger advantage might be obtained from more modest, noncompetitive forms of physical activity. Although there is some evidence that the relative risk of exercising diminishes with aging (Vuori, 1995), this is probably because the elderly are less likely to take very strenuous forms of activity, and such data cannot necessarily be extrapolated to WMG participation.

How far is a combination of excitement and all-out effort likely to provoke a heart attack relative to less intense and emotionally-charged forms of physical activity (Shephard, 1974; Shephard, 1995; Vuori, 1995)? In the very old, there is some evidence that any form of regular exercise may shorten rather than lengthen the lifespan (Linsted et al., 1991). There still may be a gain in terms of quality-adjusted life expectancy, but even if this is not the case, it is possible that Masters participants (and others) would prefer to die slightly earlier while participating in a favorite form of physical activity, rather than endure a painful terminal illness in the bleak surroundings of a nursing home. The critical age at which such a choice might prevail would depend on the social and

188

roy j. shephard

economic circumstances of the senior; but from a policy perspective, it would be useful to know the desired average survival, and factors modifying this.

Quality-adjusted life expectancy

A senior's quality of life may be compromised by various forms of chronic ill health, but for most people a much more important issue is the preservation of adequate function to live independently (Paterson et al., 2004; Shephard, 1997; Shephard, in press). Potentially modifiable determinants of independence include maximal aerobic power, the strength of key muscles, and the flexibility at major joints. Continued participation in Masters sport may play a role in keeping some or all of these physiological characteristics above the critical thresholds needed to undertake the activities of daily living independently (Paterson et al., 2004; Shephard, 1997; Shephard, in press; see Hawkins, Chapter 4), although it remains to be clarified how far self-selection contributes to the advantage of the athlete (see Tanaka, Chapter 3), and the specificity required for most types of sport training seems likely to lessen the effectiveness of such preparation relative to more general physical activity programs.

Understanding of the aging process

At first inspection, the very consistent changes in various types of athletic records with age would seem to offer a unique and precise insight into rates of human aging (see Stones, Chapter 2). However, in practice, a number of considerations complicate the interpretation of existing data. In many events, the outcome does not depend on a single physiological variable such as maximal oxygen intake. A decrease in weekly training schedules and an accumulation of body fat seem to move in parallel with age (Shephard et al., 1995), so that the performance of an older person is compromised by a greater body mass and a lower level of training. Practice and experience counterbalance the inherent effects of aging to differing amounts in different sports (see Baker & Schorer, Chapter 5). Finally, in the older age groups, a smaller subject pool and a lower level of competition reduce performance relative to the records observed in a young adult. Nevertheless, there remains the potential to exploit this approach in a variety of sports, limiting data to a series of observations made longitudinally on subjects who maintain training schedules and body composition relatively constant.

There have been many attempts to compare the course of the aging process in athletes and in sedentary individuals, collating cross-sectional and longitudinal data for various physiological measurements such as maximal oxygen intake and peak muscle force (Shephard et al., 1995), but the reported values have frequently been compromised by changes in the training schedules of the athletes due to age or specific interventions. In general, the deterioration in maximal oxygen intake and muscle force seems to proceed a little more slowly in an active population, although there is a need to obtain more reliable data (Shephard et al., 1995; also see Chapters 3 and 4 of this volume). If such a benefit is confirmed, the question again arises whether the advantage is maximized through participation in Masters sport or through more general forms of physical activity.

Maintenance of skill and dexterity are other desiderata in older adults, although it is unclear how far Masters sport involvement helps in this regard. Performance in sports that require considerable skill peaks at a later age than it does in disciplines dependent largely upon physiological characteristics, and skill-dependent activities also seem conserved later into old age. Possibly, as in other forms of physical activity such as piano playing and typing, Masters sport participation may help in the process, extensive practice (Krampe & Ericsson, 1996), and the development of compensating skills (Salthouse, 1984), serving to counteract the decline in physiological attributes. In the future, this may be explored directly by longitudinal studies of Masters participants that divide performance at any given age between its skill and physiological components.

Another significant facet of the aging process is a progressive loss of cognitive function. There are tantalizing suggestions, many derived from animal rather than human studies, that regular physical activity may delay such losses, whether through an increased synthesis of neurotrophic factors (Cotman & Berchtold, 2002), the genesis of additional neurons (van Praag et al., 1999) and an increased production of anti-oxidants (Soman et al., 1995), or more simply through increased social contacts and a broadening of interests. The friendships formed and the travel associated with competing in Masters sport seem likely to be helpful in this regard (Shephard et al., 1995), although there remains a need for careful studies comparing Masters participation with other forms of physical activity and mental gymnastics (such as solving crossword puzzles).

Injuries and infections

A final issue is the potential risks of musculoskeletal injury and upper respiratory infection associated with intensive training and competition.

roy j. shephard

Injuries

Some studies of young adults have suggested that the social costs of medical services and injury-related work loss are greater for sports participants than for the sedentary population (Nicholl et al., 1991; Reijnen & Velthuijsen, 1989). Likewise, many exercise programs for middle-aged patients have been plagued by a high incidence of musculoskeletal injuries; in some cases, as many as half of the class sustained a serious injury within the first six months of conditioning (Mann et al., 1969; Pate & Macera, 1994). The risk rises with both age and the rate of progression of the exercise program; there is thus a need to compare the risk of injury at various ages between Masters Athletes and those engaged in more pedestrian programs. In a more long-term context, the strengthening of muscles and bones, a quickening of reflexes, and an improvement of balance seem likely to reduce the risks of injury in the elderly, although the question again arises whether a more general exercise program would have advantages relative to preparation for a specific Masters sport.

Infections

The increased risk of upper respiratory infections associated with extreme training and competition has been well documented for younger individuals (Nieman, 2000; Shephard, 2000). Given the deterioration of immune function with aging (Shinkai et al., 1998), the likelihood of such problems is increased in the senior. Again, the risk is probably greater for Masters involvement than for more general types of physical activity program, although many Masters competitors are aware of a weekly training ceiling beyond which the risk of upper respiratory infection increases (Shephard et al., 1995).

CONCLUSIONS

Much more research is required before we can assert categorically either that Masters sport has a major positive impact on community health, or that this benefit is greater than could be obtained through the advocacy of simpler noncompetitive forms of physical activity. Nevertheless, investigators have claimed a wide variety of advantages for those participating in Masters sport. Reported benefits include enhanced personal survival and quality of life, amelioration of various aspects of the aging process, and the setting of a positive example to inactive peers. Certainly, Masters sport has a positive impact on most of those who participate, encouraging their regular physical activity, providing increased social contacts, and offering them a substantial broadening of interests. Adverse consequences also seem remarkably few. Governments should thus welcome Masters sport as a valuable tactic in campaigns to enhance the fitness of their populations.

REFERENCES

Cotman, C.W., & Berchtold, N.C. (2002). Exercise: a behavioral intervention to enhance brain health and plasticity. *Trends in Neuroscience*, 25, 295–301.

Karvonen, M.J., Klemola, H., Virkajarvi, J., & Kekkonen, A. (1974). Longevity of endurance skiers. *Medicine & Science in Sports*, 6, 49–51.

Kavanagh, T., Lindley, L.J., Shephard, R.J., & Campbell, R. (1988a). Health and socio-demographic characteristics of the Masters competitor. *Annals of Sports Medicine*, 4, 55–64.

Kavanagh, T., Mertens, D.J., Matosevic, V., Shephard, R.J., & Evans, B. (1988b). Health and aging of Masters athletes. *Clinical Journal of Sports Medicine*, 1, 72–88.

Krampe, R.T., & Ericsson, K.A. (1996). Maintaining excellence: Deliberate practice and elite performance in young and older pianists. *Journal of Experimental Psychology: General*, 125, 331–359.

Levy, B.R., & Banaji, M.R. (2002). Implicit ageism. In T.D. Nelson (Ed.), *Stereotyping and prejudice against older persons* (pp. 27–48). Cambridge, MA: MIT Press.

Linsted, K.D., Tonstad, S., & Kuzma, W.J. (1991). Self-report of physical activity and patterns of mortality in Seventh-Day Adventist men. *Journal of Clinical Epidemiology*, 44, 355–364.

Lockwood, P., Chasteen, A., & Wong, C. (2005). Age and regulatory focus determine preferences for health-related role models. *Psychology and Aging*, 20, 376–389.

Mann, G.V., Garrett, H.L., Murray, H., & Billings, F.T. (1969). Exercise to prevent coronary heart disease. *American Journal of Medicine*, 46, 12–27.

Montoye, H.J., Van Huss, W.D., Olson, H.W., Pierson, W.O., & Hudec, A.J. (1957). *The longevity and morbidity of college athletes*. Lansing, MI: Phi Epsilon Kappa Fraternity, Michigan State University.

Nicholl, J.P., Coleman, P., & Williams, B.T. (1991). Pilot study of the epidemiology of sports injuries and exercise-related morbidity. *British Journal of Sports Medicine*, 25, 61–66.

Nieman, D.C. (2000). Special feature for the Olympics: Effects of exercise on the immune system: exercise effects on systemic immunity. *Immunology & Cell Biology*, 78, 496–501.

Ory, M., Hoffman, M.K., Hawkins, M., Sanner, B., & Mockenhaupt, R. (2003). Challenging aging stereotypes: Strategies for creating a more active society. *American Journal of Preventive Medicine*, 25, 164–171.

Paffenbarger, R.S., Hyde, R.T., Wing, A.L., Lee, I.-M., & Kampert, J.B. (1994). Some inter-relationships of physical activity, physical fitness health and longevity. In C. Bouchard, R.J. Shephard, & T. Stephens (Eds.), *Physical activity, fitness and health* (pp. 119–133). Champaign, IL: Human Kinetics.

Pate, R., & Macera, C. (1994). Risks of exercising: Musculoskeletal injuries. In C. Bouchard, R. J. Shephard, & T. Stephens (Eds.), *Physical activity, fitness and health* (pp. 1008–1018). Champaign, IL, Human Kinetics.

Paterson, D.H., Govindasamy, D., Vidmar, M., Cunningham, D.A., & Koval, J.J. (2004). Longitudinal study of determinants of dependence in an elderly population. *Journal of the American Geriatric Society*, 52, 1632–1638.

Reijnen, J., & Velthuijsen, J.W. (1989). Economic aspects of health through sport. In Conference proceedings, *Economic impact of sport in Europe*. Lilleshall UK, November 1989, 1–31.

Salthouse, T. (1984). Effects of age and skill in typing. *Journal of Experimental Psychology: General*, 113, 345–371.

Sarna, S., & Kaprio, J. (1994). Life expectancy of former athletes. *Sports Medicine*, 17, 149–151.

Shephard, R.J. (1974). Exercise and sudden death: A significant hazard of exercise? *British Journal of Sports Medicine*, 8, 101–110.

Shephard, R.J. (1995). Exercise and sudden death: An overview. *Sport Science Review*, 4 (2), 1–13.

Shephard, R.J. (1996). Habitual physical activity and quality of life. *Quest*, 48, 354–365.

Shephard, R.J. (1997). *Aging, physical activity and health*. Champaign, IL: Human Kinetics.

Shephard, R.J. (2000). Special feature for the Olympics: effects of exercise on the immune system: overview of the epidemiology of exercise immunology. *Immunology & Cell Biology*, 78, 485–495.

Shephard, R.J. (in press). Maximal oxygen intake and independence in old age *British Journal of Sports Medicine*. doi:10.1136/bjsm.2007.044800.

Shephard, R.J., Kavanagh, T., Mertens, D.J., Qureshi, S., & Clark, M. (1995). Personal health benefits of Master's competition. *British Journal of Sports Medicine*, 29, 35–40.

Shinkai, S., Konishi, M., & Shephard, R.J. (1998). Aging and immune response to exercise. *Canadian Journal of Physiology & Pharmacology*, 76, 562–572.

Soman, S.M., Ravi, R., & Rybak, L.P. (1995). Effect of exercise training on antioxidant system in brain regions of rat. *Pharmacology, Biochemistry and Behavior*, 50, 635–639.

van Praag, H., Kempermann, G., & Gage, F.H.(1999). Running increases cell proliferation and neurogenesis in the adult mouse dentate gyrus. *Nature Neurosciences*, 2, 203–205.

Vuori, I. (1995). Exercise and sudden cardiac death: Effects of age and type of activity. *Sports Sciences Review*, 4(2), 46–84.

CONTRIBUTORS

Joseph Baker
School of Kinesiology & Health Science
York University
Canada

Rylee A. Dionigi
School of Human Movement Studies
Charles Sturt University
Australia

James Fell
School of Human Life Sciences
University of Tasmania
Australia

Steven A. Hawkins
Department of Exercise Science and
   Sports Medicine
California Lutheran University
USA

Sean Horton
Department of Kinesiology
University of Windsor
Canada

Nikola Medic
Centre for the Built Environment and
   Health
University of Western Australia
Australia

William J. Montelpare
School of Kinesiology
Lakehead University
Canada

Jörg Schorer
Institute for Sport Science
Westfälische Wilhelms-University
   Münster
Germany

Roy J. Shephard
Professor Emeritus
University of Toronto
Canada

Michael Stones
Department of Psychology
Lakehead University
Canada

Hirofumi Tanaka
Department of Kinesiology and
   Health Education
University of Texas at Austin
USA

Patricia Weir
Department of Kinesiology
University of Windsor
Canada

Andrew Williams
School of Human Life Sciences
University of Tasmania
Australia

INDEX

199